THE
WONDER
OF
GENETICS

THE
WONDER
OF
GENETICS

The Creepy, the Curious, and the Commonplace

RICHARD V. KOWLES

Prometheus Books

Essex, Connecticut

Prometheus Books

An imprint of Globe Pequot, the trade division of
The Rowman & Littlefield Publishing Group, Inc.
4501 Forbes Blvd., Ste. 200
Lanham, MD 20706
www.rowman.com

Distributed by NATIONAL BOOK NETWORK

British Library Cataloguing in Publication Information Available

Library of Congress Cataloging-in-Publication Data

Kowles, Richard V:
    The wonder of genetics : the creepy, the curious, and the commonplace /
Richard V. Kowles.
        p. cm.
    Includes bibliographical references and index.
    ISBN 978-1-61614-214-8 (cloth)
    1.   Genetics-popular works. I. Title.

QH437.K69 2010
576.5—dc22

2010020540

ISBN 978-1-63388-946-0 (pbk.)

∞™ The paper used in this publication meets the minimum requirements of
American National Standard for Information Sciences—Permanence of Paper for
Printed Library Materials, ANSI/NISO Z39.48-1992

This book is dedicated to my wife, Rose;
our children, Doug, Greg, Debora, Jeanne, and Brian;
and their spouses, Sandy, Jim, Mike, and Eileen,
who have become part of the Kowles clan.

CONTENTS

PREFACE

The concepts of genetics permeate society everywhere. Genetic principles are the bases of many facets of everyday events experienced by everyone. The goal of *The Wonder of Genetics* is to demonstrate how the discipline of genetics plays a role in these societal aspects that people encounter in their everyday lives. In meeting that goal, it is hoped that the reader will also become better informed about how genetics actually works. Much of the public is excited about genetics, or at least they find it interesting. Other people, however, are still gravely concerned about the potential applications of the discoveries being made in the fields of genetics and its closely related discipline, molecular biology. Almost every week, newspapers and magazines are reporting some kind of genetic breakthrough. Geneticists are constantly announcing their discoveries to the world, such as genes that they have isolated, or genes that they have identified for a particular trait such as alcoholism, aging, a disease, or a deviant behavior.

People cannot escape from questions and concerns about their everyday experiences. Some couples wonder why they had two or three consecutive miscarriages. Some couples want to know about the genetic risks of their son or daughter marrying and having chil-dredn with his or her first cousin. Some people are concerned about radiation exposure from a mammogram being more harmful than helpful. Many couples have had a Down syndrome child and are concerned about the chances that their next child will also have Down syndrome. Other people wonder whether it is safe to eat genetically modified tomatoes or other such foods. Some men are skeptical about whether they are really the biological fathers of a newborn child. Members of mixed marriages almost always want to know what to expect regarding the skin color of their offspring. Almost everyone would like to see a fountain of youth discovered. And, of course, just about everyone is worried about the onset of cancer. What is the common basis underlying these questions and concerns? It's genetics.

Many people have other questions that may not be so immedi-ately dire. They may question the development and use of genetic advances, the existence of free will, stem cell research, and the cloning of organisms, especially humans. They may have been involved in serious debates over controversies such as evolution versus creationism, nature versus nurture, human gene therapy, and racial issues. The underlying source of these controversies to a large extent is also genetics. Women in the Olympics, some movies, car-toons, and many political decisions also follow a genetic trail.

The content of this book is deemed to be "softer" than many other science books aimed at the general public, but hopefully without a sacrifice of genetic accuracy. An attempt has been made to explain complex concepts in a simple, succinct manner without an overload of technical genetic jargon. Still, genetics has a vocabulary of its own, and it is impossible to ignore all of it. In addition to sim-plistic explanations, an extensive glossary is provided. The manner of writing is often in a conversational voice with the use of real examples, comparisons, and analogies wherever pertinent.

Most chapters begin with examples of everyday exposure to genetics, some of which the reader may not have realized have a genetic connection. Discussions follow to reveal the substance of the genetic connection, and also to carefully explain the concept fundamental to the topic—all in understandable terms. The book is a series of thirty-seven chapters that delve into thirty-seven facets of everyday life experiences having a genetic connection. Some of these connections are quite apparent, and some may not be very apparent. The discussions probe the genetics involved in societal problems on one hand, but also solutions on the other hand, along with applications to many walks of life. An attempt is made to place them into perspective with explanations designed for the average reader. The author strives to make clear what is known, what is opinion, what is factual information, and what is simply wishful thinking. Whatever people might be mindful of, a good chance exists that there's a genetic connection.

The author wishes to acknowledge individuals who were very helpful throughout the development of this book. Patti Reick created ten of the illustrations. Jim Cummings helped with the photographic technology of figures. Laurie Robertson and Maribeth Bedkte gave secretarial help, and Rose Kowles did proofreading. Also Steven L. Mitchell and the Prometheus Books staff were easy to work with and extremely helpful. All issues that arose were intelligently discussed in a professional manner.

CHAPTER 1

INTRODUCTION—
GENETICS EVERYWHERE

What factors contribute to make us what we are? Why are we male or female, short or tall, blonde or brunette, blue-eyed or brown-eyed? Why do we sometimes resemble our parents, siblings, aunts, or uncles? Why do we sometimes not resemble any of our relatives? Why are some people born with defects in anatomy (such as extra fingers), defects of biochemistry (such as the inability to metabolize the amino acid phenylketonuria), or defects of neural function (such as manic depression)?

Researchers in the field of genetics investigate all of these questions, and more. All living organisms, no matter how primitive or complex, are the result of the workings of genetics. Living organisms are composed of lifeless molecules. These molecules conform to all of the physical and chemical laws that describe inanimate matter. Yet living organisms possess extraordinary attributes not shown by inanimate matter. Living organisms have molecular organization, and these molecules have very specific functions. Living organisms can extract and transform energy from the environment. Living organisms have self-replication. If living organisms are composed of molecules that are intrinsically inanimate, why then does living matter differ so drastically from other matter? The answer lies in living organisms being able to replicate their genes. These units of inheritance, the genes, hold the information necessary to direct the construction of enzymes that are the protein tools active in cellular reactions and absolutely essential for life itself. These processes are the essence of genetics.

Genetics is a fascinating science dealing with the inheritance of physical and chemical characteristics. More specifically, genetics is the discipline involved in the study of variation due to heredity and the effects of environment on this variation. Heredity is the overall process of transmitting traits from parents to offspring. In other words, heredity is the tendency to be like your parents. To a large sigh of relief for some, "tendency" only means "tendency."

Almost everyone agrees that the actual birth of genetics came about with Gregor Mendel, an Austrian monk. Mendel devoted years to a study of several hereditary traits, or characters, that he observed in common garden peas. His classic paper explaining his plant hybridization experiments first appeared in journal form in 1866, although no one gave his results serious attention at that time. Mendel collected a large amount of data indicating that hereditary characteristics were due to particles of some kind. He called these particles "factors" and showed that they passed from one generation to the next according to predictable patterns. He further observed that definitive hereditary characteristics (such as short or tall peas) did not blend together into an average form (such as medium-height peas). The characteristics remained in their original form, generation after generation.

Many students of the history of genetics, however, see 1900 as the true birth date of genetics, for it was then that three biologists independently rediscovered Mendel's paper. As a result, the basic laws of heredity became known, and studies of them could then progress at a rapid pace. Yet traditional biologists were slow to accept studies of heredity as a distinct discipline. William Bateson named the field "genetics" in 1905, and it became a respectable science about 1910.

During the period from 1900 to 1910, scientists made significant contributions to genetics. In 1903, W. S. Sutton and T. Boveri independently and concurrently made the correlation between the pattern of inheritance of Mendel's factors and the behavior of chromosomes in cells. They concluded that chromosomes were the vehicles carrying these hereditary particles. William Bateson and R. C. Punnett, among others, conducted many plant hybridization experiments during this time, again with peas but with other organisms too. These researchers quickly learned that heredity was certainly more complicated than Mendel had thought. In most cases, they supported the independence of two gene pairs in cell division as set forth by Mendel, but they discovered at the same time that exceptions occurred.

By 1911, the fruit fly (*Drosophila*) had become a popular experimental organism, and Columbia University's eight-desk "fly room" had become a center of genetic discovery. The fly room was the professional home of Thomas Hunt Morgan and other notable pioneers of genetics. These researchers chose the fruit fly as an appropriate organism for genetic studies because it is small, easy to culture, and convenient to handle. Determining the sexes of these flies is straightforward, and crosses could be easily set up. The generation time is short, and resultant offspring are numerous. Fruit flies were found to have many different observable hereditary traits making possible a wide variety of investigations. Furthermore, the giant chromosomes visible in the salivary glands of *Drosophila* larvae allowed researchers to observe the detailed morphology of the chromosomes themselves.

Clearly, the history of genetics is a long series of fascinating discoveries. The science was not conceived until 1866 and only truly born in 1900; however, since the identification of the hereditary material as DNA in 1944, the discipline has grown into a giant. Today, many people are greatly excited about genetics, while others are gravely concerned about the potential applications of the discoveries that remain to be made in the field of genetics and the related field of molecular biology.

Almost every week, newspapers are reporting some kind of genetic breakthrough, especially concerning news about the discovery of specific genes. Geneticists continually announce their discoveries to the world, such as genes that they have isolated, or genes that they have identified for one trait or another such as aging, some kind of disease, or a deviant behavior. Such feats represent a tremendous amount of work, ingenuity, and accomplishment. News about every gene discovery is usually regarded as a milestone toward a possible treatment or cure for something. Newspapers would not publish this information if the public wasn't interested. Of course, some newspaper embellishment usually helps to increase this interest.

Too many people are frightened of science in general and genetics in particular. Many have a very distrustful view of scientists. It is of little wonder why this apprehension exists. Surveys of TV programs have shown that 20 percent of the time scientists are portrayed as villainous, 33 percent of the time they are failures, 5 percent of the time they kill someone, and 10 percent of the time they get killed. These are terrible impressions of science and scientists. Scientists are simply people who do experiments. Experiments can literally be beautiful with regard to a conception of ideas and the way an experiment is carried out. Not too many accomplishments can exceed the excitement of taking an idea made into a hypothesis about something scribbled on scratch paper, designing an experiment to test the hypothesis, and finally acquiring data that will add to the information of how nature works.

That heredity affects everyone's life is a gross understatement. Genes have an important impact on people. They intimately affect health, personality, parenthood, work ethics, and various social concerns—sometimes in surprising ways. We are products of heredity. Genetics is a tremendous success story, but being accepted by society has not been easy. Genetics can solve problems, and yes, genetics can create problems. The discipline is an extensive component of medicine, agriculture, and generally all of biology; and genetics is already becoming an important part of psychology. Sociology may be next, but some still oppose this connection. At any rate, critical decisions need to be made about genetic advances, decisions that indeed could affect humankind. Think about the situations, good, bad, or even downright ugly, that we encounter every day. Think also about the common denominator that exists among many of these situations. Serious concerns abound with such issues as genetic engineering, reproductive manipulation, cloning, and human genetic testing. We also think about AIDS, cancer, alcoholism, mental defects, prenatal diagnoses, consanguinity (inbreeding), developmental sexual disorders, aging, human sterility, resistance to antibiotics, and numerous hereditary diseases. This litany of everyday concerns can be extended even more by the inclusion of the stem cell research controversy, the nature versus nurture controversy, the evolution-creationism controversy, DNA testing, race issues, and numerous other topics. What is the common aspect of all these topics? It is genetics. Even many comics, cartoons, movies, and everyday events have relevancy to genetics. Genetic concepts are everywhere.

FURTHER READING

Bornstein, J., and S. Bornstein. 1979. *What Is Genetics?* New York: Julian Messner.

Kohler, R. E. 1994. *Lords of the Fly: Drosophila Genetics and the Experimental Life.* Chicago: University of Chicago Press.

Reilly, P. R. 2004. *Is It in Your Genes?* Cold Springs Harbor, NY: Cold Spring Harbor Laboratory Press.

Stubbe, H. 1972. *History of Genetics.* Cambridge, MA: MIT Press.

Sturtevant, A. H. 1965. *A History of Genetics.* New York: Harper & Row.

REFERENCES

Avery, O.T., C. M. Macleod, and M. McCarty. 1944. "Studies on the Chemical Nature of the Substances Inducing Transformation of Pneumococcal Types." *Journal of Experimental Medicine* 79: 137–58.

Bateson, W., E. Saunders, and R. C. Punnett. 1904–1908. "Experimental Studies in the Physiology of Heredity." *Reports to the Evaluation Committee of the Royal Society* 2: 1–154; 3: 1–53; 4: 1–60.

Mendel, G. 1866. "Experiments in Plant Hybridization." English translation reprinted in *Classical Papers in Genetics*, edited by J. A. Peters, 1–20. Englewood Cliffs, NJ: Prentice-Hall, 1959.

Sutton,W. S. 1903. "The Chromosomes in Heredity." *Biological Bulletin* 4: 231–51.

CHAPTER 2

MALES AND FEMALES— THERE ARE DIFFERENCES

T he twentieth century began with the knowledge that heredity was due to a substance of some kind transmitted from generation to generation according to definite basic laws. In addition, biologists had known of the existence of chromosomes since 1865. Chromosomes are easily stained minute structures visible in the nucleus of the cell by microscopy. In the early 1900s, W. S. Sutton and T. Boveri independently realized that Mendel's so-called factors, which we now call genes, behaved in the same way as the chromosomes in cells. In both cases, (1) they are found in pairs; (2) one member of each pair has a maternal origin, and the other has a paternal origin; (3) members of the pairs segregate from each other during the formation of sex cells; and (4) they form pairs again following fertilization of the egg by a sperm. Sutton and Boveri concluded that Mendel's factors were carried by chromosomes. This conclusion is now called the Chromosome Theory of Heredity. Numerous experiments since then have shown these ideas to be correct. Chromosomes indeed are the vehicles of heredity.

Chromosomes viewed with a good microscope are beautiful

structures. When in their compacted condition, chromosomes sometimes look like little bow ties (figure 2-1). These observations are
most interesting when the chromosomes are your own. You then
come to the realization that these little rod-shaped structures, forty-
six in each human cell, contain most of the genetic information necessary to orchestrate the formation of a person—namely, yourself.
When photographed and paired by size and various physical landmarks, the arrangement is called a karyotype. Staining properties
and other characteristics of chromosomes are well known to the
people who study these microscopic structures. In humans, the chromosome pairs are given numbers from 1 to 22 according to their size,
with number 1 being the longest and number 22 being the shortest.
The two sex chromosomes comprising the remaining pair are called
X and Y, without a designated number. In mammals, these latter two
chromosomes play a major role in sex determination.

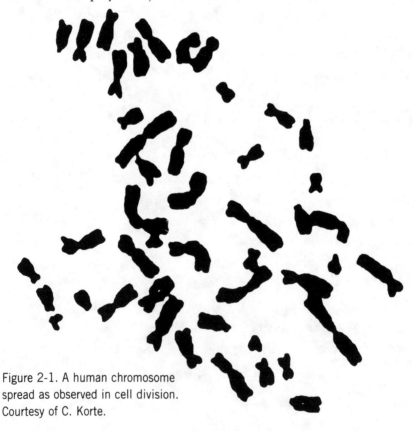

Figure 2-1. A human chromosome
spread as observed in cell division.
Courtesy of C. Korte.

Very compelling reasons exist for sex. This statement refers to genetic sex—that is, the mode of reproduction whereby gametes (sex cells) are combined by fertilization. Almost all organisms do it. Even bacteria can carry out antics similar to a sex life. The two scientists who found conjugation among bacteria (J. Lederberg and E. L. Tatum) won the Nobel Prize for their discovery. Contrasting genders among different species are not always called males and females, but two or more mating types are always necessary. Sometimes the mating types are called A and a (like in some molds), or donors and recipients (like in bacteria). Regardless of the terminology, DNA (the genetic material) from one individual forms a union with DNA from another individual. This union of DNA (fertilization) is sex in a genetic sense. In the case of animal hermaphrodites and many plants, one organism provides both gamete types. Such a reproductive mode is still considered sexual.

Not all species, however, need to have a mother and a father. Some fish, amphibians, lizards, and a few other organisms can reproduce in another way. These organisms sometimes produce eggs that develop directly into progeny without fertilization. Males are completely left out of the system. The phenomenon is called parthenogenesis, meaning virgin birth. Many a human female has probably found herself pregnant and screamed "parthenogenesis" or something equivalent, but such an event in humans has never been scientifically proven. Usually, women's stories were ultimately changed, or a lack of understanding was apparent concerning the "ins and outs" of reproduction. In some cases, the surprised woman even gave birth to a son. This latter claim of parthenogenesis is easy to dismiss since the child would, of course, have to be a female because of the chromosome situation.

Following fertilization and early development, some embryos develop into males and some develop into females. In mammals at least, the two diverse developments are due to a couple of chromosomes called X and Y. These two chromosomes are called sex chromosomes, as opposed to all other chromosomes of the cell, which are referred to as autosomes. The sex chromosomes had to be called something, and X and Y have no special meaning. The designations simply serve as names. Sex chromosomes are sometimes given different designations in other species like Z and W. In mammals, such as humans, males possess one X chromosome and one Y chromosome, and females have two X chromosomes and no Y chromosome.

Sex determination in mammals is rather straightforward, at least from a chromosome standpoint. The sex chromosomes undergo segregation to different daughter cells during the formation of gametes just as all of the other chromosomes do. The result of this segregation, assuming it is normal, is that females (XX) generate eggs that always contain one X chromosome. But males (XY) will produce two types of sex cells relative to the sex chromosomes, that is, those having an X chromosome and others having a Y chromosome (figure 2-2). Fertilization of eggs by sperm results in XX offspring (females) and XY offspring (males). The presence or absence of the Y chromosome is the paramount criterion for male or female development. Consequently, the male sex cells determine the sex of the progeny in mammals. The king who killed his wives for not giving him a son was the one responsible. He could not get his son without providing the right sperm to the egg, that is, a sperm with a Y chromosome.

The X chromosome is one of the largest of all chromosomes in humans, and it houses numerous genes. A human cannot survive without having at least one X chromosome. And there is no such thing as a female X or a male X. An X is an X is an X. An X chromosome can be passed to progeny from the male, the female, or both, but the X chromosomes have no inherent physical or gender differences. The Y chromosome, on the other hand, is the "guy" chromosome. Under normal conditions, the Y chromosome is only found in males. The Y chromosome is a runt compared to the X chromosome, and it contains a meager number of genes (maybe about fifty) compared to the X chromosome (probably a thousand or more). The Y chromosome has been heavily ridiculed, sometimes called a landfill, a genetic couch potato, a wasteland, or a genetic

Female gametes

Male gametes		X	X
	X	XX	XX
	Y	XY	XY

2 XX = females

2 XY = males

1:1 ratio

Figure 2-2. The chromosomal basis of sex determination in humans.

desert. Adding to the lack of respect for this chromosome, one can find cartoons showing the Y chromosome with the arbitrary placement of genes on it for causing spitting, reading the paper while on the john, sports news, channel flipping, selective hearing loss, and the like. Seriously, a few real genes are indeed located on the Y chromosome, and among these genes is the one absolutely essential for causing a male to develop. This master gene is called the sex-determining region on the Y (SRY). If you get the X chromosome from dad, you will be a girl. If you get the Y chromosome from dad, you will be a boy. At the very least, the presence or absence of the Y chromosome determines the sex of progeny.

The SRY gene on the Y chromosome was always kind of a Holy Grail of sex determination. For many years researchers attempted to find *the* gene on the Y chromosome that shifted development toward maleness. Tenacious scientists have found it. Of course many genes are involved in the development of the final male product, but the

SRY gene appears to instigate and regulate most of this development. Other genes have now been located on the Y chromosome. For example, some of these genes have something to do with fertility and sperm production. An interesting discovery is that "housekeeping" genes also exist on the Y chromosome. Once more, "housekeeping" genes are found on the Y chromosome. Housekeeping genes in the genetic sense are those genes that are probably expressed in all cells of the body. They are necessary for maintenance of activities that simply keep cells alive. Most housekeeping genes, therefore, are probably located on the X chromosome as well as the Y chromosome, because females do not have a Y chromosome and XX cells have to carry out the same cellular maintenance. Genetic activity is gradually being associated with the Y chromosome. It is not quite the genetic desert it was thought to be at one time.

Sex determination is actually a complex process, and it can go wrong in various ways. Babies are born each year who are difficult to identify as either male or female. Often they are intersexes with ambiguous external and/or internal sexual organs and tissues. Many children are born each year with ambiguous genitalia. More than twenty-five different kinds of intersex conditions can occur. Most of them are arbitrarily designated as being females probably because it is an easier surgery in order to make them morphologically like females. But will some of them grow up feeling like males?

True hermaphrodites, who have both ovarian and testicular tissues, are rare among live births. They usually have an ovary on one side with a few follicles, and a testis on the other side with at least a few seminiferous tubules. However, human hermaphrodites are not reproductively functional as either sex. Hermaphrodites are best described as intersexes showing sexual ambiguity, and they often lack typical sex characteristics. Most have XX chromosomes, while an XY chromosome condition is rarer. Some hermaphrodites are XX/XY mosaics; that is, some of their cells are XX and some cells are XY. The specific mechanisms that produce hermaphrodites are not well understood.

Sometimes an individual feels like a male being trapped in a female body, or like a female trapped in a male body. Simply put, these individuals think their body has the wrong plumbing. Transsexualism refers to voluntary sex changes by people with ambiguous external genitalia or simply by those who feel more comfortable as the opposite sex. The reasons are usually psychological, and most often men who wish to become women choose transsexual surgery. The transformation also involves the application of sex hormones. I once met a person who made such a drastic change. She, or he (I'll just call him/her by R. R.), was a male turned female. R. R. had quite a grip when shaking hands. R. R. gave a very intelligent and insightful talk to a large audience about the ordeal of feeling female while being trapped in a male body.

Several different levels of sex obviously exist. There is primary sex, described simply by the gonads you own; it determines which public toilet you use. The presence or absence of a Y chromosome relates to chromosomal sex. Secondary sex includes the traits associated with masculinity or femininity; for example, the presence or absence of a beard, the pitch of the voice, the size of the breasts, body contour, and so on. We also have to include the psychological level of sex, that is, the "what gender do you feel like" aspect. One's perception of sexual identity is an important issue that deserves careful analysis. Serious questions about maleness, femaleness, and the differences between the sexes, as well as about the proper treatment for psychological and physical difficulties, need to be addressed. Sexual anomalies, as a group, are not all that rare.

FURTHER READING

Bainbridge, D. 2003. *The X in Sex: How the X Chromosome Controls Our Lives.* Cambridge, MA: Harvard University Press.

Koopman, P. 1995. "The Molecular Biology of SRY and Its Role in Sex Determination in Mammals." *Reproduction, Fertility, and Development* 7: 713–22.

REFERENCES

Callahan, G. N. 2009. *Between XX and XY: Intersexuality and the Myth of Two Sexes.* Chicago: Chicago Review Press.

Lederberg, J., and E. L. Tatum. 1946. "Gene Recombination in *Escherishia coli.*" *Nature* 158: 588.

Palmer, M. S. 1989. "Sex Determining Genes." *Science Progress* 73: 245–61.

Sutton, W. S. 1903. "The Chromosomes in Heredity." *Biological Bulletin* 4: 231–51.

CHAPTER 3

FEMALES ARE GENETIC MOSAICS

A cellular difference between mammalian males and females is easily visible with an appropriate stain and a good microscope. The difference was first discovered in the nerve cells of cats in 1949. It involves the chromosomes in the nucleus of the cell that can be detected in most mammalian cells. In humans, this male versus female chromosome difference is conveniently clear in cells scraped from the lining of the inner cheek (called buccal cells), transferred to a microscope slide, and suitably stained. In cells from a female, a small deeply stained body is visible just inside the nucleus. This little structure is called the Barr body or sex chromatin. The Barr body is not observed in male cells. Although females have two X chromosomes, evidence shows that in each of their cells only one X chromosome is completely active; that is, its genes are actually churning out messages. The other X chromosome is largely inactive, and it condenses into the small discrete structure called the Barr body. Males lack Barr bodies because the one X chromosome they have in each cell needs to remain completely active.

The determination of which X chromosome will be active and

which will be inactive in a female cell occurs very early in embryo development. Substantial parts of the X chromosome are inactivated as early as sixteen days after conception (not birth). Much of the chromosome is inactive by five weeks after conception. In some cells, the inactive X chromosome is the paternal one (received from the father), and in other cells the inactive X chromosome is the maternal one (received from the mother). The inactivation of paternal versus maternal X chromosomes is thought to be random in most body tissues, differing from cell to cell during that early stage of the embryo development. But once the inactivation of the X chromosome is made, the same X chromosome (paternal or maternal) is inactivated in all of the descendants of that cell following cell division.

Generally no more gene product is synthesized from the two X chromosomes in a female cell than from the one X chromosome in the male cell, all of which is due to the inactivation process. A female who is homozygous (two copies) for an X-linked deleterious gene is usually no more affected than a male who has only one copy of the deleterious gene (designated as hemizygous). Apparently, females undergo a phenomenon called dosage compensation by means of this X chromosome inactivation. In most cases then, every person has one fully functioning X chromosome per cell regardless of how many X chromosomes are actually present in the cell.

Females can have different forms of a gene (called alleles) paired on their two X chromosomes, meaning they are heterozygous. Therefore, if a female is heterozygous for one or more genes located on the X chromosome, she is said to be mosaic relative to that gene pair. Some of her cells will express one form of the gene, while other cells of the body will express the other form of the gene (figure 3-1). Regardless of this mosaic situation, a heterozygous female will usually show the phenotype (appearance) of the dominant normal allele. Evidently, enough cells are present in the body expressing the dominant normal trait to result in normal appearance. For example, a female heterozygous for the hemophilia gene (bleeder disease), which is located on the X chromosome, will almost always have a

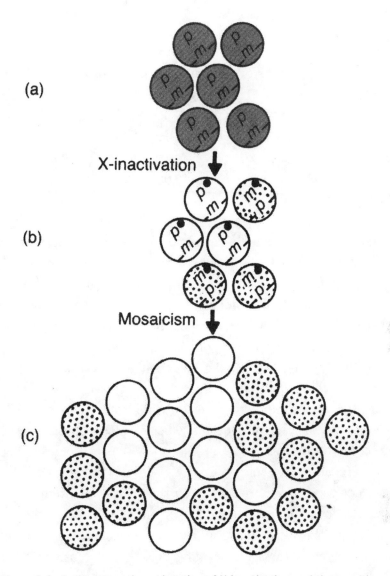

Figure 3-1. A diagrammatic explanation of X-inactivation and the resulting mosaicism. The symbols "p" and "m" indicate the paternal and maternal origins of the X chromosomes. The dark and light cells signify different gene activities in the two types of cells. (a) A small group of cells is depicted for clarity (researchers have estimated that the actual number of cells present at the time of inactivation is about one thousand to two thousand); (b) the X-inactivation event; (c) only one X chromosome remains fully active in each cell of a mosaic embryo. Eventually a full-grown mosaic individual will develop.

normal blood-clotting system. Extremely rare exceptions do exist because it is remotely possible for a heterozygous female to have mostly cells with the normal X chromosome inactivated simply due to random inactivation. However, the probability of such an occurrence is very slim. At any rate, such rare females would express the recessive condition even though they have a dominant allele that would normally mask the recessive allele. Such rare females are called manifesting heterozygotes. The bottom line, however, is that females should not lie awake at night worrying about being X chromosome mosaics.

Anhidrotic dysplasia is another genetic syndrome due to a recessive allele located on the X chromosome. The genetic defect reduces the number of sweat glands in the skin. A heterozygous female will then show a mosaic situation in which parts of her body will sweat normally, and other parts of her body will not sweat at all because of the syndrome. This mosaic situation has been confirmed by literally mapping the regions of the body of such individuals for their ability to sweat, with the results completely dependent on which X chromosome has been inactivated in the various parts of the body. Males with an X chromosome containing the anhidrotic dysplasia allele will have a completely sweat-defective body. Since males have only one X chromosome, any gene on that one X chromosome will be expressed regardless of being recessive or dominant. An inability to sweat is not good because perspiration is a crucial requirement for cooling the body. Once a college basketball team had just such an individual on its roster. The young man was an excellent player, but he could only play for a few minutes at a time because of overheating and consequently turning red as a beet. Some unsympathetic members on the opposing team thought that he might literally blow up in their faces.

Anytime you see a calico cat, wager with anyone lacking genetic savvy that the cat is a female. Calico cats have orange and black fur color interspersed throughout their coat. Chase one down, roll it over, and look closely at the right body parts. You will win. This color pattern is another example of a mosaic situation in which the two different alleles (orange and black) are located on separate X chromosomes of the female cat.

FURTHER READING

Hanna-Alava, A. 1960. "Genetic Mosaics." *Scientific American* 202: 118–20.
Russell, L. B., ed. 1978. *Genetic Mosaics and Chimeras in Mammals.* New York: Plenum.

REFERENCES

Barr, M. L., and E. G. Bertram. 1949. "A Morphological Distinction between Neurons of the Male and Female, and the Behavior of Nucleolar Satellite during Accelerated Nucleoprotein Synthesis." *Nature* 163: 676–77.

Lyon, M. F. 1961. "Gene Action in the X Chromosome of the Mouse (*Mus musculus* L.)." *Nature* 190: 372–73.

———. 1962. "Sex Chromatin and Gene Action in the Mammalian X Chromosome." *American Journal of Human Genetics* 14: 135–48.

———. 1988. "The William Allan Memorial Award Address: X-Chromosome Inactivation and the Location and Expression of X-Linked Genes." *American Journal of Human Genetics* 42: 8–16.

CHAPTER 4

ANYONE EXACTLY LIKE YOU?

Almost all of us have at one time or another experienced someone approaching us with a greeting by a name not ours. Evidently you closely resembled some other person. One begins to think that an uncanny resemblance between you and this other phantom person must exist.

Let's explore the biological possibility of two people being nearly genetically identical—without being identical twins. To do even a simplistic analysis, one must understand meiosis, the process by which we churn out sex cells. Chromosomes come in pairs, and the two chromosomes of each pair are called homologues. These two chromosomes of a pair are very similar to each other, but not identical. Homologous chromosomes are the same size, have the same shape, have the same structural landmarks in the same location, and harbor genes at the same sites up and down the chromosome. However, a pair of genes residing on the pair of chromosomes can have different forms; that is, they can be allelic to each other. Therefore, genes constituting a pair on homologous chromosomes do not have to be identical. They can be allelic; for example, A blood type versus B blood type, normal pigmentation versus albinism, and so on.

In some cases, the alleles are identical, making the pair homozygous, and in other cases, the alleles are different from each other, making them heterozygous. In either case, the two chromosomes are nonetheless homologous, and they will pair during meiosis. Assuming that you do not have an anomaly regarding chromosome numbers, twenty-three pairs of homologous chromosomes exist in each of your cells.

Two main benefits result from the meiotic process. First, this type of cell division keeps the chromosome number constant from generation to generation. Even small changes in chromosome number can have disastrous effects on the development of an individual. The second benefit is that the process results in a great amount of genetic variation among our sex cells. Meiosis separates the members of a pair of chromosomes into sets made up of single chromosomes from each pair. Fertilization (conception to many people) then combines these sets of single chromosomes from each of the two parents into new pairs, making many new combinations of alleles. These events are very important for generating genetic diversity.

Meiosis is a long, complex, and exciting cellular process. The event includes two successive chromosome segregations, that is, distributions of chromosomes during two cell divisions (meiosis-1 and meiosis-2). To begin, all of the chromosomes replicate into sister chromatids. Then, the members of homologous chromosome pairs physically embrace each other. In meiosis-1, the homologues, each made up of two identical sister chromatids due to the replication event, segregate into two daughter cells. In meiosis-2, the sister chromatids segregate into other daughter cells. The result is that four daughter cells develop into sperm in the male or an egg and three polar bodies in the female. The polar bodies are nonfunctional and eventually degrade (figure 4-1).

A little arithmetic can show the immense diversity realized as the result of meiosis. Since chromosomes exist in pairs and humans

Figure 4-1. Meiosis is illustrated in a cell in which only two pairs of chromosomes are shown for clarity. To aid in following the movements of the chromosomes, "A" and "a" mark one pair of chromosomes, and "B" and "b" mark the other pair.

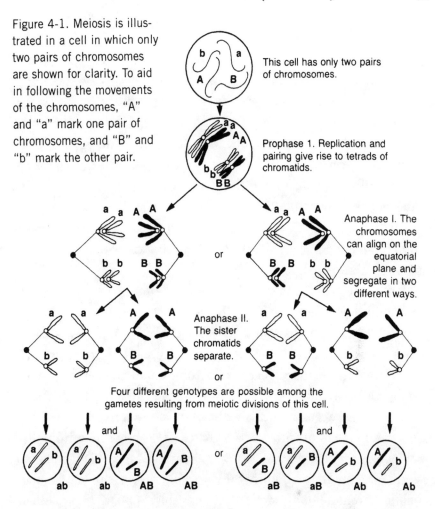

This cell has only two pairs of chromosomes.

Prophase 1. Replication and pairing give rise to tetrads of chromatids.

Anaphase I. The chromosomes can align on the equatorial plane and segregate in two different ways.

Anaphase II. The sister chromatids separate.

Four different genotypes are possible among the gametes resulting from meiotic divisions of this cell.

have 23 pairs, any individual is capable of producing 2^{23} different combinations of the chromosomes as a result of meiosis. This calculation means that you have to multiply 2 by itself a total of 23 times. A calculator will easily show that 2^{23} equals 8,388,608. And some of these pairs could show different alleles; that is, the two genes making up the pair could have different forms and consequently relate to different expressions. If you can produce 8,388,608 different chromosome combinations, so can your mating partner. The number of combinations now becomes 2^{46}, or 8,388,608 multiplied by

8,388,608, resulting in the possibility of 70,368,744,000,000 different chromosome combinations in the zygotes (fertilized eggs). And this staggering figure does not take into account that chromosomes can also exchange parts to further increase diversity.

The consideration of inbred organisms is a different situation. With inbreds, all of the genes located on a particular chromosome are virtually identical to the genes on the other homologous chromosome of the pair; in other words, they are completely homozygous. Therefore, it makes no difference how the chromosomes assort during meiosis; the gametes will all be identical. Hence, all of the progeny will look alike except for subtle differences due to environmental influences. If genetic differences occur, the parents were not entirely inbred.

Could it be possible that two sibs (brother-brother or sister-sister) born at different times be genetically identical? Or on the other hand, is it possible that two sibs would have no genes in common with each other by descent? The answer is that both situations are *technically* possible. Meiosis dictates that one chromosome from each pair is given by each parent to the origin of the zygote. Genetically identical sibs would only occur if the same batch of chromosomes from one parent would meet up with the same batch of chromosomes from the other parent on two separate baby-making occasions. But randomness and probability predict that the possibility of such a happening is infinitely remote and practically impossible. A good guess is that this situation has never happened at any time in the history of humankind. The same is probably true for two completely different batches of chromosomes coming together on two different occasions to form two completely different progeny (figure 4-2). On the average, however, sibs who are not identical twins have about 50 percent of their genes in common with each other.

The possibility of being extremely different from each other is reminiscent of the film *Twins* starring Danny DeVito and Arnold Schwarzenegger. The situation is unlikely, even as fraternal twins, but again remotely possible—very remotely. However, the story in

(a) **(b)**

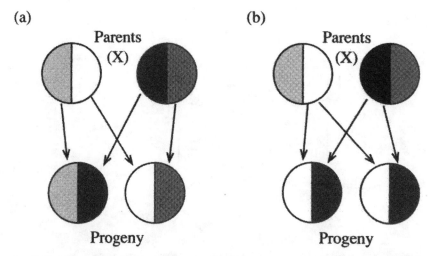

Figure 4-2. A diagrammatic illustration of siblings being (a) completely different from a genetic standpoint; and (b) completely identical from a genetic standpoint. Both situations are only theoretically possible but not at all probable. Courtesy of Patti Reick.

Twins had Danny and Arnold as identical twins with one of them receiving the "good" genes and the other the "bad" genes because of some bizarre experiment (really bizarre). One can always sense fiction when it is encountered.

Recombination is a phenomenon that leads to even more gene shuffling. During early meiosis, the chromosomes swap bits of genetic material in a process known as recombination. This exchange of genetic material, due to a physical event of chromosomes crossing over, takes place between homologous chromosomes and leads to a new combination of genes; hence, it is called recombination (figure 4-3). These chromosome antics mean an establishment of more combinations of genes in our sex cells. Crossovers and the resultant recombination of genes do not happen by accident. Rather, these events constitute a precise design for generating sex cells, most of which will have their own genetic identity. Together with segregation and random assortment, the mechanism of crossing over additionally contributes to genetic variation among our sex cells.

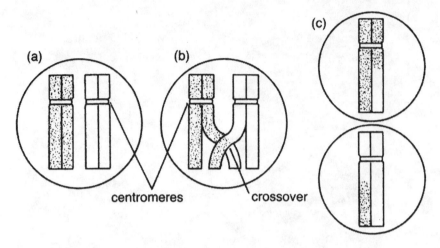

Figure 4-3. Chromosome crossing over. (a) Replication into sister chromatids and pairing of homologous chromosomes produce a tetrad (four) of
chromatids; (b) exchange of segments between nonsister chromatids;
(c) the result of the crossover event leads to new combinations of genes.

So theoretically, how many different kinds of sex cells can an individual make? First, some assumptions need to be made. Assume that we have 25,000 gene pairs and that hypothetically we are het-erozygous for about 10,000 of them. Again, crossing over between homozygous alleles changes nothing. Under these circumstances, the answer becomes $2^{10,000}$. Imagine the size of that number whereby one would multiply 2 by itself 10,000 times. You may have also seen the film *Dave*. Someone off the street took the place of an ailing president of the United States because he looked exactly like the president, and nobody detected the deception—except the real president's wife. And she realized the difference not by how he looked, but rather by how he acted. Cute film story, but it is probably not possible. So again, some place on this earth someone supposedly exists who is geneti-cally identical to you? No, that person probably does not exist.

FURTHER READING

Therman, E. 1980. *Human Chromosomes: Structure, Behavior, and Effects.* New York: Springer-Verlag.

REFERENCES

John, B., and K. R. Lewis. 1976. "The Meiotic Mechanism." In *Oxford Biology Reader* 65, edited by J. J. Head. London: Oxford University Press.

Lewis, K. R., and B. John. 1972. *The Matter of Mendelian Heredity.* New York: John Wiley & Sons.

Oldroyd, D. 1984. "Gregor Mendel: Founding-Father of Modern Genetics." *Endeavour* 8: 29–31.

Sutton, W. S. 1903. "The Chromosomes in Heredity." *Biological Bulletin* 4: 231–51.

CHAPTER 5

PLAYING GENETIC ROULETTE

Approximately five thousand (or more) traits of the human have already been assigned to a genetic etiology. Undoubtedly many more will be added to the list. The mutant form of many of these traits is detrimental, that is, downright bad stuff. Some of these conditions, therefore, can rightly be called diseases. They are genetic disorders. Like other diseases, these conditions are also transmitted, but vertically rather than horizontally. Most people probably think of diseases as only being due to bacteria, viruses, fungi, various parasites, and other bad little creatures. Many such diseases are communicable and transmitted from person to person within the population, that is, horizontal transmission. Genetic diseases, however, cannot be imparted to each other in the present population. But we can transmit them from one generation to another generation, that is, vertical transmission. Genetic factors, however, may actually be of some importance in the origin of almost all pathogenesis.

At a college some time ago, a general education course was taught called Heredity and Society. It was a course designed to study and discuss numerous genetic diseases and their mode of inheritance, among many other human genetic situations. During this time, the college employed a health consultant to determine the extent of health classes being taught in the curriculum of the college. After the "expert completed the study," the final report came out without any mention of the Heredity and Society course among the various courses that included health issues.

Data concerning genetic ill health for the United States are constantly compiled, and they are not pretty. People with inherited diseases occupy approximately 25 percent of all hospital beds, and 30 percent of all children admitted into hospitals are dealing with genetic diseases. Of all childhood deaths, 42 percent are caused by genetic defects, and a substantial proportion of these deaths are due to the absence of appropriate treatment. Between 5 and 10 percent of the population has an inherited defect, and in 2 to 3 percent of the population, the defect is severe. Conservatively, these figures relate to about fifteen million people in the United States. It is believed that 20 percent of all heart attacks occur because of one or more gene defects playing a role. About 2 to 3 percent of the US population is mentally challenged (IQ below 70), and as many as 80 percent of these mental conditions might have a genetic basis.

The bad news is that many different severe genetic defects exist. The good news is that individually most of them are rare. Taken together, these numerical findings lead to the overall 2 to 3 percent occurrence in the population. More bad news is that everyone, on the average, probably has three to five deleterious genes within his or her repertoire of genetic material. More good news: most of these deleterious gene forms are recessive; that is, the other allelic gene of the pair is normal and dominant. Dominance is designated to the trait expressed in an individual (normal or abnormal) when the gene pair is in a heterozygous condition (both allelic forms present).

Although the normal form of a gene is often the dominant one, genetic dominance does not infer an advantage. Huntington disease, a state of mental deterioration, is due to a dominant gene. Polydactyly, extra fingers and/or toes, is due to a dominant gene. Brachydactyly, characterized by short, very thick fingers, is due to a dominant gene. One form of ichthyosis, a scaly fishlike skin, is due to a dominant gene. Achondroplasia, a form of dwarfism, is also due to a dominant gene. And many more could be listed. None of these traits can be thought about in the context of providing any kind of an advantage.

To take part in this genetic roulette activity, all you need to do is have children. Since many of these genetic defects are recessive, normal dominant genes (alleles) mask the genes responsible for the defects, and most of the time individuals don't know which recessive genes are lurking among their genetic machinery. They can truly be considered to be "genes in the closet." However, if you happen to have one of these gene versions, you would not want your mating partner to have the same recessive gene. But you usually have no control over this situation. Most of the time, parents do not know whether either of them is carrying that particular culprit gene. So having children is really playing genetic roulette.

The chances that your partner is heterozygous (a carrier) for the same problem gene that you have is not very high, which is more good news. But of course, it is possible that you and your partner could both be carriers for the gene in question. Some of these gene forms are not that rare. Under these conditions, the probability that your newborn will be homozygous for the recessive gene is one out of four. Both genes of the pair need to be the recessive deleterious form in order for the genetic disease to express in the newborn.

A common saying about probability is that it has no memory. This assertion, which is generally true, means that each newborn has the same chance of having the genetic defect, that is, one out of four. This probability, however, pertains strictly to genetic diseases caused by a pair of recessive genes. Could a parental couple under these conditions have three successive affected children? Yes! Could a parental couple under these conditions have six successive children, none of them affected? Yes! When a parental couple has an affected child, they do not then have three unaffected children coming to them. Some parents have actually believed that when visiting a genetic counselor. Once again recall that probability has no memory. People having children are playing genetic roulette.

The theoretical ratio, therefore, when the parents are both heterozygous for a recessive gene is three unaffected to one affected. Such a ratio indicates that the trait is inherited as Mendelian; that is, the trait is caused by one pair of alleles. Human albinism is just such a recessive genetic defect. The problem is not as severe as many other recessive disorders, but it still has some serious disadvantages. Affected individuals lack the brown-black pigment melanin found in skin, hair, and eyes. Albino individuals are extremely fair-skinned, and the iris of their eyes is pinkish since the blood vessels of the eye are not masked by melanin. Visual problems, sensitivity to sunlight, and an increased risk of skin cancer often accompany albinism. The incidence of albinism is reportedly to be one in ten thousand to twenty thousand live births.

Scientists fully understand only a few of the many genetic defects caused by recessive alleles. Cystic fibrosis affects about one in two thousand Caucasians. The responsible allele is the most widespread recessive defect found in the United States, with about one in twenty Americans believed to be carriers (heterozygous). In children with the syndrome, the respiratory tract fills with thick mucus, the lungs become severely infected, secondary infections become common, gastrointestinal problems often occur, and much salt is lost in perspiration. Affected individuals usually require constant care.

Phenylketonuria (PKU) is routinely detected and treated in newborns. Although the condition is fairly rare, affecting about one in ten thousand to fourteen thousand newborns, the screening program for the syndrome is considered an overwhelming medical success. The genetic disease is due to a biochemical anomaly that causes the child to be mentally challenged, sometimes severely, and have other neurological problems such as convulsions. A simple test performed shortly after birth reveals whether the blood contains an elevated level of the amino acid phenylalanine. A high level of phenylalanine, in turn, will sometimes indicate PKU, in which case a strict diet low in phenylalanine is immediately prescribed. The positive results of a prompt treatment are dramatic. Clear differences in the development of the mental capabilities in treated versus untreated individuals have been easily shown. All infants in the United States are now tested for PKU shortly after birth.

Tay-Sachs disease is rare except in certain Jewish populations. Among the Ashkenazic Jews of Eastern Europe, its incidence is one in 3,600, and one person in 30 is a carrier for the gene. In the general non-Jewish population, the incidence of Tay-Sachs is only one in 360,000. The symptoms of Tay-Sachs disease appear within one to six months after birth. These signs include the child showing a mentally challenged situation, with the possibility of blindness, deafness, and paralysis. The child literally wastes away and usually dies between the ages of two and four years. There is no cure, and even with the best medical care, the child will most certainly die by age five. Tay-Sachs is a metabolic disorder in which lipids (gangliosides) accumulate in the tissues, especially in the central nervous system. One of the enzymes normally responsible for lipid breakdown is missing in the tissues of Tay-Sachs victims. Carriers can be detected by blood tests. However, it makes no difference that we know the exact genetic and biochemical cause of Tay-Sachs disease. Nothing can be done about it. Usually, enzyme deficiencies cannot be rectified by simply supplying the missing enzyme to the victim. Taken orally, the enzyme would be digested in the intestines like other proteins. Injection into the bloodstream will not allow the enzyme to get into all of the body cells either. The molecule is too large to get around.

Xeroderma pigmentosum is rare, but it has been extensively studied because of its close association with cancers in skin areas repeatedly exposed to sunlight. Xeroderma pigmentosum is initially characterized by pigmentation abnormalities, dry scaly skin, and possibly some scarring. A hypersensitivity to the ultraviolet component of sunlight causes the eruption of many skin tumors, largely because the cells of these persons lack the ability to repair the breaks that often occur in the DNA molecules within the cells. These chromosome breaks can occur spontaneously and in response to agents such as ultraviolet light and background radiation.

Alkaptonuria has sometimes been called the "black urine" disease because the urine of alkaptonuria individuals tends to turn dark when exposed to light and air. This situation must be quite a surprise if the parents are not aware of the problem before they start changing diapers. But the disease is more serious than just black urine. One could always conceal embarrassing moments in the toilet. The disease also appears to be linked with the onset of arthritis later in life, sometimes very severe. Alkaptonuria is another one of the classic metabolic disorders in which the cause is biochemically well defined. When the body metabolizes the amino acid tyrosine, the substance is normally converted to homogentisic acid, among other substances. Homogentisic acid, in turn, is broken down by the enzyme homogentisic acid oxidase, but in persons with alkaptonuria, this enzyme is missing. Consequently, homogentisic acid accumulates, and much of it ends up in body tissues, joints, and urine.

This brief discussion only touches upon a few of the many genetic syndromes. About 4,994 other genetic diseases could be included. Add up all of the probabilities of having children with one of these genetic maladies, and the overall probability becomes about 3 percent, although each of them individually is a rare occurrence. When we have children, it is truly a game of genetic roulette.

FURTHER READING

McKusick, V. A. 1998. *Mendelian Inheritance in Man: A Catalog of Human Genes and Genetic Disorders.* Baltimore, MD: Johns Hopkins University Press.
Reilly, P. R. 2000. *Abraham Lincoln's DNA and Other Adventures in Genetics.* Cold Spring Harbor, NY: Cold Springs Harbor Laboratory Press.

REFERENCES

Okada, S., and J. S. O'Brien. 1969. "Tay-Sachs Disease: Generalized Absence of a Beta-D-N-acetylhexosaminidase Component." *Science* 165: 698–700.

Wills, C. 1970. "Genetic Load." *Scientific American* 222: 98–107.

CHAPTER 6

THE IMPORTANCE OF MAMA

Not all of our DNA is located in the nucleus of the cell. Some DNA, which is also made up of genes, has its residence in the mitochondria of plant and animal cells and also in the chloroplasts of plant cells. This DNA is arranged in extremely tiny closed circles. Mitochondria (organelles) are small structures in the cytoplasm (outside of the nucleus) of the cell. These organelles provide most of the molecules involved in generating energy; hence, they can loosely be called the powerhouses of the cell, and for that matter, the organism. Chloroplasts of plant cells, another type of organelle, are instrumental in photosynthesis (food making), and they also contain DNA. Consequently, mitochondria and chloroplasts have their own little genome (assembly of genes)—probably about forty to fifty different genes, but there are many copies of each of these genes owing to multiple circles of DNA in every organelle.

Traits expressed due to these genes in the cytoplasm are always passed to the offspring by the maternal parent. Thus, such heredity is called maternal inheritance or cytoplasmic inheritance. This type of inheritance is considered non-Mendelian. Maternal inheritance changes the rules because it does make a difference as to who is the maternal parent and who is the paternal parent.

The concept underlying maternal inheritance deals with the biology of the fertilization event. When a sperm enters an egg, only the head of the sperm, fully packed with DNA of the male, actually gets into the egg. The mitochondria in animals, for example, are situated just behind the head in the neck of the sperm. Of course the mitochondria supply the energy so that the sperm can act like a little motorboat in their pursuit of the coveted egg. But the mitochondria do not enter the egg, meaning that their genes belonging to the male are not going to represent the male in the newly formed zygote. Since it is not a perfect biological situation (few are), occasionally a few mitochondria from the male might slip in and crash the party. In such cases, the chemical machinery in the egg will degrade them. No way are they allowed admittance. Sorry males, but sometimes this maternal inheritance process can be a good thing—for example, when the male mitochondrial genes happen to be deleterious because of mutation events.

Figure 6-1 illustrates how maternal inheritance works when contrasting gene types (alleles) exist between the male and the female. Note that the hypothetical traits, symbolized by a shaded background versus a clear background, occur in the progeny according to which mitochondrial gene the female owns, and not the mitochondrial gene of the male. Symbols for nuclear genes (alleles) are also presented to show that they too are not a factor with regard to the maternal inheritance of the cytoplasm. The transmission of cytoplasmic genes completely depends upon the maternal contribution.

In humans, at least thirty-seven genes are known to be located on the DNA of the mitochondria. Some of these genes code for molecules necessary in the generation of cellular messages. Other genes code for RNA messages per se, which, in turn, will generate polypeptides in order to make certain proteins. Cytoplasmic genes are essentially the same as nuclear genes in their composition, that is, made up of the amazing DNA molecules. Consequently, these

Diploid species

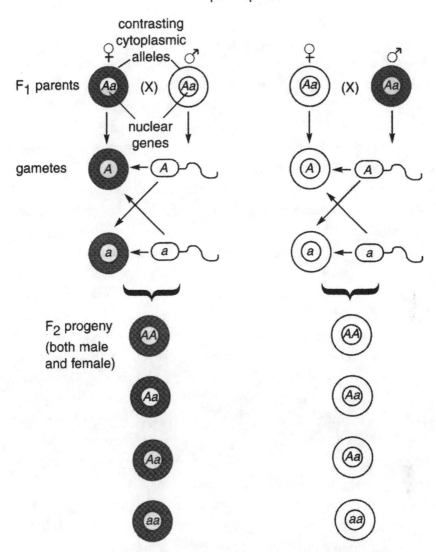

Figure 6-1. Crosses showing the non-Mendelian relationships observed when contrasting alleles exist in the cytoplasm. Note that the nuclear alleles segregate and recombine in a Mendelian manner, but transmission of the cytoplasmic alleles completely depends upon the maternal contribution. From R. V. Kowles, *Solving Problems in Genetics* (New York: Springer, 2001). Reprinted with permission.

genes can also undergo mutations and various DNA rearrangements, all of which can lead to defective physiological functions for the individual containing them.

Defective mitochondrial genes in humans can bring about a number of genetic maladies. Optic neuropathy is one such example. This blindness is a strange condition that occurs by the time a person reaches twenty years old. The syndrome can sometimes occur as a very sudden loss of vision. Today a person can see and tomorrow the person cannot see. Other examples include a form of epilepsy, a form of deafness, a form of diabetes, Kearns-Sayre syndrome (paralysis of eye muscles, dementia, and seizures), and several neuromuscular diseases. The known number of such syndromes is rapidly increasing with the greater amount of research attention being given to the problem. The gene composition of the mitochondrial DNA and the number of copies of each gene can greatly vary from cell to cell in a person, and even within individual cells. This particular situation is called heteroplasmy, and the concept is responsible for the great amount of variation in the degree of severity observed in these cytoplasmic genetic syndromes. They can range from mild to lethal.

An interesting and kind of clever use of cytoplasmic inheritance has been devised in the breeding of hybrid corn. Genes in the chloroplast DNA follow a maternal inheritance pattern just like the mitochondrial DNA; that is, these genes are only transmitted to the progeny via the maternal parent. Chloroplasts cannot be transmitted into the egg since they are not even part of the male sperm in plants. One of these genes has an allele that causes male sterility in progeny. Therefore, some strains of corn were discovered in which normal pollen development did not occur. These plants were male sterile, but they still produced normal ears that could be pollinated and fertilized by other corn plants. The inheritance of this type of gamete lethality is cytoplasmic, meaning that the cytoplasm of the zygote primarily comes from the maternal parent's egg.

This maternal inheritance situation became valuable in the production of hybrid corn, where the objective was to cross two different strains of corn with each other but to disallow self-fertilization of at least one of them. It was accomplished by planting rows of male-fertile plants next to rows of male-sterile plants. The male-fertile corn would produce pollen, and the male-sterile corn would produce a normal ear for an effective cross. The male-fertile plants could, of course, self-pollinate and would not be hybrids. They were simply used for other purposes like feed for cattle. The resultant hybrid plant, however, would be male sterile, and seed would not be produced under these conditions. Farmers would not be very happy with this turn of events. They would only be growing corncobs without any corn kernels. The dilemma was erased with the discovery of a nuclear gene, the *Rf* allele (restorer of fertility), which can restore fertility regardless of the cytoplasm. The male-fertile plants need to be homozygous for the *Rf* allele. Since this restorer allele is dominant to its recessive allele, the hybrid will pro-

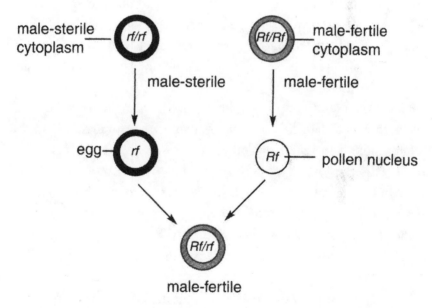

Figure 6-2. Cytoplasmic inheritance of male sterility in maize is additionally dependent upon the restorer of fertility alleles (*Rf* and *rf*); consequently, male-sterile cytoplasm can still result in fertile plants. From R. V. Kowles, *Solving Problems in Genetics* (New York: Springer, 2001). Reprinted with permission.

duce pollen for seed production. This use of genetics was very popular because it obviated the need for plant-breeding companies to make laborious controlled crosses by hand to obtain hybrid seed. Figure 6-2 illustrates how this clever scheme works.

Almost 80 to 90 percent of the United States corn crop in 1970 consisted of varieties containing this particular type of cellular cytoplasm. These varieties gave excellent yields, and the seed was economical to produce, so farmers preferred them. But the varieties with this cytoplasm became susceptible to a race of corn blight (*Helminthosporium maydis*). Before 1970, most strains of corn were quite resistant to this plant disease. However, the fungus had apparently mutated enough over time to make this heavily used strain of corn vulnerable. The blight destroyed 15 percent of the nation's overall corn crop and half of the crop in some heavily infested regions. Because of this vulnerability, the corn blight brought the country to the brink of a major crisis. The carbohydrates (starches and sugars) of corn are used in a myriad of products. United States science and industry managed a rapid recovery from disaster, but the new corn blight race also caused some long-term changes in breeding hybrids.

The country is now less fond of such genetic uniformity, and new corn varieties quickly became available. Also, some readers may be old enough to recall the other way to make corn plants male sterile. It is the task of going up and down the rows of corn and removing the tassels by hand before they shed pollen, literally emasculating them. The author recalls doing this handiwork as a poor youngster enthused about making twenty-five cents per hour. For some plant breeding companies, detasseling of corn by hand has returned.

One should not confuse maternal inheritance due to cytoplasmic genes with maternal inheritance due to nuclear genes. The latter form of inheritance (called maternal effect) also exists, and it is another way to have a maternal influence on the initial qualities of the progeny. Maternally derived substances, including proteins and various molecular messages synthesized in the egg, may exert an influence on the early development of the embryo independently of the nuclear genes of the embryo. These factors are mostly involved in the process of normal embryo development, and they are not self-perpetuating. Rather, the molecules eventually degrade and disappear. Such factors will then be replaced by substances dictated by the genes of the existing nuclei of the embryo (genes contributed by both parents). Males should not be discontent about all of this maternal inheritance stuff because they should realize that they contributed half of the nuclear genes. And nuclear genes are responsible for the bulk of the genetics of the progeny. Still, the importance of mama's genetics is striking in so many ways.

REFERENCES

Albertsen, M. C., and R. L. Phillips. 1981. "Developmental Cytology of 13 Genetic Male Sterile Loci in Maize." *Canadian Journal of Genetics and Cytology* 23: 195–208.

Gracen, V. E. 1982. "Types and Availability of Male Sterile Cytoplasms." In *Maize Biological Research*, edited by W. F. Sheridan, 221–23. Charlottesville, VA: Plant Molecular Biology Association.

Singh, R. J. 2000. *Plant Cytogenetics*. Boca Raton, FL: CRC Press, 61–62.

Sprague, G. F., and J. W. Dudley, eds. 1988. *Corn and Corn Improvement*. Madison, WI: American Society of Agronomy, 704–10.

CHAPTER 7

THOSE AWFUL SIDESHOWS

A cruel way to treat human beings is to parade them in carnival and circus sideshows, advertised as "human curiosities," or "human oddities," or even "freaks." These people, of course, had absolutely no choice in how they would appear when they were born. Generally, circus goers would pay a small amount of money for an admission ticket to go into tents to gape at these various unfortunate individuals. Most of the sideshows are now history; however, a few of them may still be in operation.

Almost all of these people were born with severe heritable disorders due to recessive or dominant mutations. Some of the disorders were the result of developmental mistakes. Such malformations in development are described as teratogenic errors. *Teratogenesis* is a Greek word that literally means the "origin of monsters."

The best known of these genetic situations is probably neurofibromatosis, also known as the "elephant man" disorder. The mutation rate for neurofibromatosis is about one in ten thousand, among the highest mutation rates calculated for humans. Much variability, however, exists in the expression of this disorder. Symptoms may range from a few brown spots on the body to the severe disorder shown in the film *The Elephant Man*. The film was based on the plight of a man in London, John Merrick, who had an extreme case of neurofibromatosis. Large tumor-like growths on his face and body caused drastic disfigurement. John Merrick was badly mistreated during much of his life as a carnival sideshow attraction. Knobby Man, who appeared in other sideshows, may also have been afflicted with neurofibromatosis.

Porcupine Man had ichthyosis hystrix. His skin was very thick, scaly, discolored, and covered with long bristle-like hairs. This person was exhibited throughout Europe. Interestingly, Porcupine Man actually fathered six children. Alligator Boy was probably another severe case of one type of ichthyosis.

Seal Boy didn't have arms; rather, he had only flipper-like appendages. This condition is aplasia of the limbs and was probably the genetic disorder called phocomelia. In addition to the spontaneous occurrence of this disorder, an environmental mimic of it also exists. This version of aplasia of the limbs is due to pregnant women taking the drug thalidomide while pregnant in order to combat morning sickness, especially between twenty-seven and thirty-three days after conception. The developmental error is actually believed to be due to a breakdown product of thalidomide. Tests of the drug prior to human use were conducted on rats and showed negative results. However, unknown before many pregnant women were already taking the drug was the fact that it does cause aplasia of the limbs in rabbits and monkeys and unfortunately humans.

Hermaphrodites were another big draw in the sideshow business. Two types of hermaphrodites can occur. True hermaphrodites, who are quite rare, contain both ovarian and testicular tissue, but neither tissue is functional. Pseudo-hermaphrodites, probably those found in sideshows, appear to have gonads representing both sexes, that is, a penis and a vagina. This situation is another anatomical irregularity due to developmental errors. Some people liked to enter the tent and gawk at and even make fun of the hermaphrodite.

Many other people with a severe heritable syndrome or developmental mistake found their way into the sideshow tents of the circus. The gangly giant might have had Marfan syndrome. The dwarfs could have been the result of a number of genetic syndromes, most notably achondroplasia. Elastic woman could pull her skin out as much as six inches and probably had Ehlers-Danlo syndrome. Hypertrichosis lanuginosa was the dominant genetic mutation probably causing the "bear woman," the "bearded lady," and the "lion-faced boy." These individuals were literally covered with hair. Some hypertrichosis is also due to a recessive gene. Leopard Girl had a mutation called piebald. Persons with extra arms or legs due to developmental errors were always big hits in these awful displays. This condition is known as heteradelphian, and it is extremely rare.

In 1907, the salary for some of these individuals was as high as $500 per week, which was an excellent wage for that period of time. Yet most of them wished that they had not been born. A reporter once interviewed a small group of these sideshow people, and all of them expressed the desire that they would rather have been aborted. Ridiculing humans suffering from genetic defects is behavior at its worse. Sideshows were exploiting genetic mishaps.

FURTHER READING

Lenz, W., and K. Knapp. 1962. "Thalidomide Embryopathy." *Archives of Environmental Health* 5: 100–105.

Montagu, A. 1979. *The Elephant Man.* New York: E. P. Dutton.

REFERENCES

Karp, L. E. 1981. "The Elephant Man." *American Journal of Medical Genetics* 10: 1–3.

Seidman, L. A., and N. Warren. 2002. "Frances Kelsey & Thalidomide in the US: A Case Study Relating to Pharmaceutical Regulations." *American Biology Teacher* 64: 495–500.

CHAPTER 8

WHY SO MANY MISCARRIAGES?

In our desire to have children, what can go wrong? Actually, a lot! In ordinary terms, miscarriage means a failure to reach a proper ending. In a genetic sense, it means spontaneous abortion. Many abortions, but not all, have an underlying genetic cause. Mutations can cause spontaneous abortions. Mutations refer to changes in the DNA that makes up our chromosomes, that is, any kind of changes. Small changes in the DNA molecule are called point mutations, while large structural changes in the chromosome structures are sometimes called gross mutations. Should these changes occur in the cells developing into sex cells (the germ line), or in the sex cells themselves, it could possibly be trouble for the subsequent embryo.

Spontaneous mutations are often due to the gross mutations, that is, the chromosome aberrations. These chromosome abnormalities are structural rearrangements of the chromosome(s). In other cases, the potential abortion problem can occur in sex cells because these cells have the wrong number of chromosomes. An argument can be made that spontaneous abortions constitute a genetic screen. Simply explained, many genetic mistakes are actually culled out as embryos before they become live births.

Mutations, point or gross, are heritable if they occur in the germ line. We pass copies of our DNA to the zygote (fertilized egg) and subsequent embryo regardless of what might have happened to the DNA, and a lot can happen to the DNA. Mutations can occur spontaneously or they can be induced by a variety of agents. Simply put, spontaneous mutations are those not artificially induced, meaning they just happen. We increase the odds of getting mutations by exposing ourselves to radiation, sunlight (containing ultraviolet light), cigarette smoke, asbestos, benzene, PCBs (polychlorinated biphenyls), arsenic, mercury compounds, and many other chemicals.

Our cells, and the DNA found within each of them, are constantly being bombarded by background radiation, medical radiation, human-made radiation, ultraviolet light, thousands of chemicals, and possibly other agents, some of which are not yet within our knowledge. We refer to them as environmental insults to our biological systems, or we just plain call them "hits." We are continually taking these genetic hits all day long, every day. Therefore, maintenance of the precious structure of our DNA is one of the primary goals of our biological system other than reproduction. This maintenance task is indeed a difficult undertaking, but our cellular repair mechanisms come to the rescue. So one should always be thankful for having good working repair mechanisms.

It is believed that most hits, potential point mutations and chromosome breakage, are repaired as long as we have these good repair mechanisms. Some data indicate, however, that caffeine, at least in high quantities, might interfere with the proper functioning of our repair systems. The author recalls some years ago having lunch with one of the premier geneticists of the time. When asked about what was new in the world of genetics, he discussed research that showed caffeine as possibly interfering with the body's repair mechanisms. In fact, he went on to state the data were so strong in this regard that if he was scheduled to have some x-rays, he would refrain from drinking any coffee previous to his x-ray appointment. While he related this dire concern, he smoked cigarettes nonstop throughout the entire lunch, even while eating. Premier scientist or not, a conversation should have followed with regard to what constitutes a hazard.

Large groups of genes, together with some protein molecules, form the structures we call chromosomes. Chromosomes are essentially DNA molecules, and they are present throughout the life of the cells, except for red blood cells in mammals. Chromosomes are easiest to observe during mitosis and meiosis, that is, during cell division processes. When cells are appropriately treated, placed on a microscope slide, and flattened with a little pressure, the chromosomes tend to spread into a nice single thin layer, allowing for accurate counting and analysis. Photographs and computer analyses of chromosomes spread in this manner, allow researchers to sort and arrange the chromosomes by size, banding patterns, and other landmarks into a karyotype (figure 8-1).

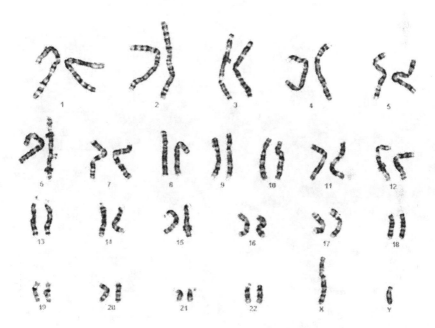

Figure 8-1. A karyotype of human chromosomes displayed by one of the banding procedures. The karyotype shows all forty-six chromosomes arranged by size. In this case, the chromosomes are from a male (X and Y chromosomes are present). Courtesy of Mayo Clinic Cytogenetics Laboratory.

Technical advances in analyzing chromosomes now allow researchers and medical personnel to recognize the presence of chromosome abnormalities in a precise way. Chromosome problems generally fall into two major categories: (1) individuals may have the wrong number of chromosomes; or (2) a particular chromosome or combination of chromosomes may have some kind of structural rearrangement.

Chromosomes can break, and when they snap into two or more

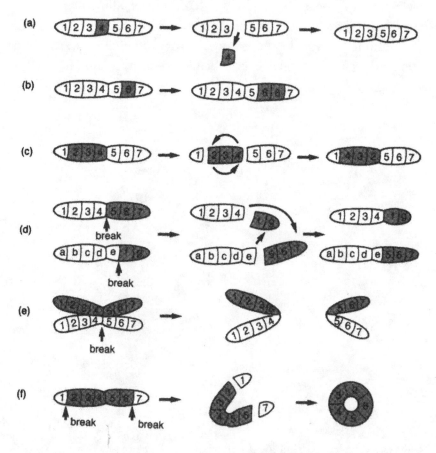

Figure 8-2. Some types of chromosome aberrations: (a) deficiency (or deletion); (b) duplication; (c) inversion; (d) reciprocal translocation; (e) isochromosomes; (f) ring chromosome.

parts, any of three outcomes may follow. The broken parts may simply be repaired back into the original form of the chromosome, that is, restored in an accurate way. This restoration event may be the most frequent outcome following chromosome mishaps. Such a repair event, of course, is good news. However, the repair mechanisms of the cell may occasionally place the broken DNA pieces together in an erroneous way; hence, the result is a chromosome rearrangement of some kind (figure 8-2). This repair event is not good news. Third, the fractured chromosome material may not be repaired at all; rather, the pieces just remain as chromosomal carnage. More bad news!

Following some aberration events, all of the chromosomal material is still available to the cell. An exception, for example, would be a lost segment of a chromosome resulting in a deficiency (also called a deletion). And a duplication would have twice as much of a particular chromosome region than normal. Regardless, even with the right amount of DNA in most of these rearrangements, grave consequences can still occur. If a chromosome gets broken, the point of fracture might occur within an important gene, causing that gene to be nonfunctional. Also, some of these chromosome aberrations will result in sex cells having an abnormal amount of chromosomal material owing to chromosome segregation during the production of sex cells, such as deficiencies, duplications, or both. The fertilized egg should have two copies of the genetic material, one copy from you and one copy from your mate. As humans, our developmental well-being does not tolerate deviations from the "two copies" fundamental basis very well. A high percentage of embryos with chromosomal deficiencies and/or duplications probably abort at some point during the gestation period, many of them very early in the developmental process.

Another serious problem can occur during the production of gametes (sex cells). Sometimes the resultant gametes will have too many chromosomes, or on the other hand, too few. Thus, when such a gamete combines with the other gamete to bring about fertilization, the resultant embryo will have a lousy beginning. That first cell, the zygote, will have too many or too few chromosomes; therefore, every cell of the developing embryo that follows due to cell division will now have this same chromosome predicament. Having one or a few chromosomes too many or too few is called aneuploidy. Having one or more entire sets of chromosomes too many is called polyploidy. These different ploidy levels, as they are called, are depicted in figure 8-3.

It is imperative that we have two doses of all the genetic information and nothing more or less. As previously mentioned, the development of a human does not tolerate deviations from this

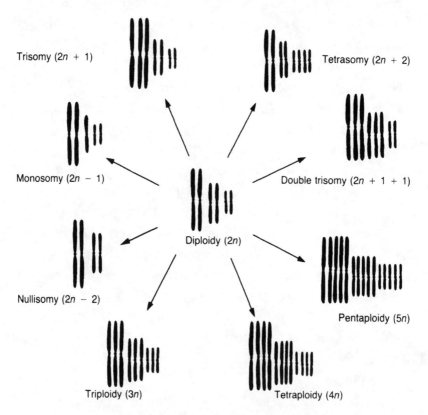

Figure 8-3. Some of the abnormal variations in chromosome number that can occur in organisms are shown. For clarity, the normal number of chromosomes in this example is shown as three pairs of chromosomes rather than the forty-six normally found in human cells.

chromosomal "two-ism." The genetic concept involved is that gene balance is very necessary. Most deviations from gene balance will abort. Other than Down syndrome (three chromosome 21s), very few examples exist whereby a trisomy, three of any particular chromosome, will allow the individual to develop all the way to a live birth. The exceptions are trisomy-13 (Patau syndrome), trisomy-18 (Edwards syndrome), and very rarely trisomy-22. That is it as far as live births are concerned. However, even these children are born with severe mental and physical abnormalities, and most of them die at a very early age. Three of any of the other chromosomes (1 through 22) can and do occur, but such trisomies will always result in a spontaneous abortion.

An abnormal distribution of chromosomes during cell division is one of the faulty mechanisms leading to aneuploidy and polyploidy. The failure of a correct chromosome distribution is called nondisjunction. Nondisjunction can occur in mitosis (division of body cells), or in meiosis (cell division for gamete production). Nondisjunction can occur in the male, the female, or both members of a parental couple. In principle, whatever chromosome distribution one can hypothetically design on scratch paper can probably occur within the cells of an organism.

Some simple arithmetic will exemplify the abortion situation. Considering all pregnancies, 15 percent end up as spontaneous abortions (miscarriages), meaning they are not an elected abortion. When researchers study the chromosome condition in these aborted embryos, approximately 50 percent of them show a chromosome abnormality large enough to be detected with ordinary microscopy. This means that at the very least 7.5 percent of all conceptions have a chromosome irregularity of some kind.

$$.15 \times .50 = .075 = 7.5 \text{ percent}$$

In addition, one in two hundred live births (0.5 percent) have been shown to have a chromosome irregularity. Adding 0.5 percent to 7.5

percent results in *at least* 8 percent of all conceptions with a chromosome problem.

.075 + .005 = .08 = 8 percent

This frequency is indisputable because there is excellent technology to observe chromosomes, and scientists certainly know how to count. But there is little doubt that this frequency is at the low end of the real situation. Researchers probably miss some of the very small aberrations that cannot be detected by ordinary microscopic means. An even greater factor left out of this calculation is that many spontaneous abortions occur so early that no one even knows that they occurred. The briefly pregnant woman sometimes might not know that she was pregnant and that she had an abortion. It is believed that some abortions may occur at the very first cell division after conception, or at least very soon after that event.

Researchers, therefore, tend to believe that the actual frequency of chromosome abnormalities among all conceptions is anywhere from 20 to 60 percent, with the majority of estimates given at approximately 30 to 35 percent. Not such a perfect chromosome world exists as some would like to believe. The probability of our being here alive and well is actually rather low. Once more, most geneticists use another description for all of these spontaneous abortions, that is, a genetic screen. In other words, most of the embryos with chromosome abnormalities are screened out by spontaneous abortion without becoming live births. Conservatively, if 7.5 percent of spontaneous abortions have a chromosome abnormality (abnormalities that can be easily visualized), and if 0.5 percent of all live births have a chromosome abnormality, one can understand that a conception under these aberrant chromosome conditions is fifteen times more likely to abort than to develop into a live birth.

.075 / .005 = 15-fold

From the standpoint of live births, the aberrant chromosome condition is very serious from both physical and mental standpoints. Most of the genetic screening takes place very early in embryo development. Spontaneous abortion seems to be a natural event in the world of reproduction. Spontaneous abortion prevents a lot of mental problems, severe physical problems, and early-age lethality. Most parents have children without chromosome abnormalities because 199 out of every 200 live births (99.5 percent) will have normal chromosomes. This good news, however, is only because the genetic screen is very effective.

FURTHER READING

Fuchs, F. 1980. "Genetic Amniocentesis." *Scientific American* 242: 47–53.
Therman, E., and M. Susman. 1993. *Human Chromosomes: Structure, Behavior, and Effects.* New York: Springer-Verlag.

REFERENCES

Biggers, J. D. 1981. "When Does Life Begin?" *Sciences* 21: 10–14.
Harnden, D. G. 1996. "Early Studies on Human Chromosomes." *BioEssays* 18: 163–68.
Paris Conference. 1971. "Standardization in Human Cytogenetics." *Birth Defects: Original Article Series* 8 (1972).
Tjio, H. J., and A. Levan. 1956. "The Chromosome Number of Man." *Hereditas* 42: 1–6.

CHAPTER 9

RADIATION— FRIEND OR FOE?

T he answer to this question, of course, is that radiation is both friend *and* foe. X-ray technology is a wonderful way to look into the body without a scalpel. Other radiation techniques also have shown great success in medical diagnostic applications. In addition, medical treatment with radiation is one of the common ways to fight disease, especially cancers. But another aspect of radioactivity that exists is the foe part.

Ultraviolet (UV) light can also be mutagenic; that is, exposure to this particular range of wavelengths can cause gene mutations. The effects and dangers of ultraviolet light, however, are often underestimated by people. Types of light are characterized by their wavelengths. The shorter the wavelength, the better it can penetrate a bodily tissue, and the more dangerous the light source becomes. Wavelengths are usually measured and related in nanometers (nm). A nanometer is only one billionth of a meter. Most of the visible light resulting in the spectrum of colors from blue to red shows a range from about 400 to 600 nm. Ultraviolet lights have shorter wavelengths than the visible spectrum and their range is usually classified as follows:

260 to 280 nm is called UVC
280 to 320 nm is called UVB
320 to 390 nm is called UVA

About 8 to 9 percent of the total energy emitted by the sun is made up of wavelengths less than 400 nm, mostly UVA. Most people have been badly sunburned at one time or another during their lifetime. Guilty individuals can thank excessive exposure to UVA for those episodes. Worse than a sunburn is the risk of skin cancer. A fairly good correlation exists between the incidence of skin cancer and the amount of exposure to sunlight. Correlation is a statistic that shows whether an association exists between two different variables that can be measured in some way. In this case, there is an association between the frequency of skin cancer and the amount of sunlight exposure. The UV component of sunlight not only injures cells, and sometimes kills them, but this form of radiation can also cause mutations, and mutations can sometimes activate cancer. Now one can understand why transfer hoods used by microbiologists working with bacteria are often equipped with UV light to prevent contamination; why surgery rooms are often treated with UV lights to keep everything sterile; and why the World Health Organization (WHO) has advocated that developing countries fill bottles with drinking water and leave them in sunlight for five or six hours to disinfect the water. UV light has a mutagenic effect on the microbes, and many of the resultant mutations are lethal to the microbes.

Skin cancer has unmistakably increased over the years. About one in fifteen hundred persons developed skin cancer in 1930. In the past decade, the incidence of skin cancer has been more like one in seventy-five people. This type of cancer is one of the fastest increasing of all the different cancers. Another significant increase is lung cancer in females. For many years now, it has been acceptable for females to smoke cigarettes anyplace and anytime, even in *public*. As one of the commercials from not too long ago pointed out, "You have come a long way, baby." Yes indeed! But all lung cancer in females, and males for that matter, is not due to smoking.

Many scientists believe the increase in skin cancer is due to the loss of ozone in the atmosphere. The ozone of the upper atmosphere normally screens out much of the UV light from the sun. The loss of this layer means that we are now getting exposed to more UVA and UVB than in past years. Risk estimates indicate that a 10 percent reduction in the ozone level would lead to a 30 to 32 percent increase in the incidence of squamous cell carcinoma. This cancer is a malignancy of the epithelium, that is, the tissue covering body surfaces. There would also be about a 22 to 23 percent increase in the incidence of basal cell carcinoma, which involves a deeper layer of cells in the skin. These cells rarely metastasize (migrate to other tissues). Some skin cancer is not overly serious. The melanoma form is serious. This type of cancer is the malignant neoplasm occurring in skin cells that produce melanin, which is the pigment of our skin. Melanoma, however, can be cured with early detection and surgery, resulting in a fairly high success rate of about 80 percent. Melanoma has an especially high frequency in persons with the disorder xeroderma pigmentosum. This genetic disorder is due to a single recessive gene that must reside in the homozygous condition in persons who have the disorder. Researchers are confident that these affected people have a faulty DNA repair mechanism. The gene in question becomes damaged in the skin cells. Consequently, affected people's cells cannot repair the gene, and melanoma is the result. These people need to stay out of the sunlight.

In the past, we were very concerned with ultraviolet light C, but not so much with ultraviolet lights B and A, and for good reason. UVC has a wavelength around 260 nm. That is exactly the wavelength absorbed best by our precious DNA. Numerous experiments have shown that ultraviolet light at 260 nm is absolutely the most effective for creating mutations in DNA, not counting ionizing radiation. Today, however, we are also becoming increasingly wary of UVB and UVA. A particular range of these wavelengths (310 to 340 nm) is called near UV. Mutations and cancer can also occur from these wavelengths, although at a lower rate of incidence than those associated with UVC. So how do we get exposed to UVB and UVA light? As already mentioned, some of those wavelengths are found in ordinary sunlight. A very small amount of UV light can also be found in fluorescent lighting and of course suntan booths. There is no such thing as a safe tan. Any level of tan indicates some level of damage to skin cells. In addition to mutation and cancer, too much sunlight can cause premature wrinkling of the skin. To paraphrase a statement from an unknown source, a good chance exists that the browned beauties of today will turn into wrinkled prunes at a not-so-later time.

Again from an unknown source, the author once heard the following perceptive statement. We worry all the time about dark skies because each year tornadoes, lightning, and floods kill about 225 of us on the average in the United States. But sunny skies kill about 8,000 of us every year due to serious skin cancer. Does this mean that the frequency of deaths due to skin melanoma is greater in Alabama, Texas, and Florida than in Wisconsin, Montana, and Washington? Yes, it does. Data have also shown these relationships.

Ionizing radiation is another threat to the stability of our genes. This radiation is more energetic and penetrating than UV light. If this kind of radiation strikes an important molecule, which is called a target, serious consequences can occur. Such radiation is capable of effecting ionization; that is, it can literally knock an electron out of its usual residence in an atom. Groups of atoms make up the important molecules of our body. Electrons are those minuscule negatively charged particles normally swirling around the nucleus of an atom. So even if the targets are initially missed, these renegade electrons can move within living tissue, imparting their energy to other atoms so as to produce other ionizations. This sort of event going on in your cells is not good because some of these subsequent ionizations could also affect your DNA molecules. Once again, when your DNA undergoes a change, you have a mutation.

The major types of ionizing radiation that are constantly bombarding us are x-rays, gamma rays, and cosmic rays. The absorption of radiation in living tissue is usually measured in units called rads or rems. For starters, you need to know that about 300 to 450 rems applied at the same time to a group of humans would kill approximately half of the group. This is called the lethal dose 50 (LD_{50}), meaning the dose necessary to render death to half of the population being exposed. It is kind of an ugly parameter since you have to kill half of some group of organisms to determine the LD_{50}. We can easily do such experiments with mice and rats, although many people are uncomfortable with that practice too. At any rate, mice and rats usually show an LD_{50} of about 600 to 700 rems. It appears that they are tougher than we are when it comes to resistance to radiation. The range of 300 to 450 rems for humans is our best estimate based upon laboratory accidents, medical miscalculations, atomic bomb casualties, and other such pertinent data.

Damage from high doses of radiation is obvious, resulting in severe burns, radiation sickness, and even death. But how concerned should we be about lower doses? People are all subjected to a certain amount of background radiation. Can background radiation cause mutations and even cancer? First, consider the amount of radiation that people are exposed to over the course of a year. Throughout this analysis, keep the numbers in perspective; that is, the LD_{50} for humans is about 300 to 450 rems. The allowable dosage to the public as set down by the United States Nuclear Regulatory Commission is 0.1 rem per year, not counting a person's particular background conditions and medical exposures. Changing rems to millirems (1/1000th of a rem) means the allowable dosage per year beyond background and medical sources would be only 100 millirems. Presenting background exposures in millirem units is more convenient because the amount of radiation being considered is relatively low.

A number of factors determine the extent of background radiation. People living in Denver (elevation about 5,200 feet) are exposed to more cosmic radiation than people living in Baltimore (elevation about 20 feet)—67 millirems compared to 28 millirems. Topography can also make a difference because of different terrestrial conditions relative to the soil and rock in the area. Calculations have been made that show a person gets an additional 14 millirems per year by living in a brick house rather than a wooden house (bricks are more radioactive than wood). Even our bodies are radioactive, mostly because of the radioactive forms of potassium and strontium absorbed into our tissues, especially our bones. One might want to quit hugging other people. Someone actually calculated that sleeping with another person increases your background radiation exposure by 0.1 millirem (0.1 millirem, not rem) per year. Even if correct, most people probably find the practice worth the *terrible* risk.

Many radiation biologists regard radon to be the greatest source of background radiation. Radon is a colorless, odorless gas that seeps through the earth and rocks as a result of the decay of naturally occurring uranium and radium. The gas can lodge in the lungs and cause lung cancer. Carcinogenic effects of radon are beyond dispute. Since radon is a gas, it can seep into homes through cracks in the foundation, the basement floor, and other extremely small structural openings. Without good ventilation, the gas can then build up in the home. The Environmental Protection Agency advocates that homes should have no more than 4 picoCurries per liter of air (a measurement of how many radioactive disintegrations of radon occur every second in a liter of air). If your home approaches 20 picoCurries per liter, reconstruction measures are probably in order. If the measurement approaches 100 picoCurries per liter, you might want to move into a motel as quickly as possible. Should you have a house tested for radon before signing a purchase agreement, usually for a large amount of money (or actually any amount)? Many home buyers are now requesting radon information before buying. Radon causes mutations, and mutations can cause cancer.

In addition to natural radiation, consumer products are all around us that are also emitting radiation: dials on some watches, some types of jewelry, cigarette smoke, color television, airport inspections, and even smoke detectors. However, smoke detectors are a definite plus when risks are weighed against benefits because of the lives these units actually save. And color TVs are now constructed in such a way that essentially makes them safe. Most of these radiation sources result in minuscule amounts, collectively exposing people to about 5 to 13 millirems per year. Medical radiation, on the other hand, can be more substantial. The average amount of medical radiation received per person per year is estimated to be about 53 to 72 millirems. But one must fully understand averages. Many people are subjected to no medical radiation per year. This means that many people then are subjected to much more medical radiation than 53 to 72 millirems per year. Averages can be deceiving.

We can also include a tiny bit of exposure due to radiation fallout that is still hanging around from those atomic bomb tests in the 1950s and 1960s, the nuclear power plants, and radioactive wastes stored here and there. These latter sources add up to only another 0.06 millirems per year as long as we don't have an accident or experience sabotage. The addition of all these sources brings the total radiation to about 260 millirems per year for the average individual (0.26 rem). These numbers have been extracted from the *National Council on Radiation Protection and Measurements: Ionizing Radiation Exposure of the Population of the United States*. To place these numbers into still another perspective, the calculations of B. L. Cohen show that 100 millirems has the same risk as smoking fifteen cigarettes, driving a car five hundred miles, or rock climbing for fifteen minutes.

Abundant data show that high doses of radiation can be mutagenic and carcinogenic. But how about very low doses? Since each of us receives an average of about 260 to 360 millirems of radiation per year, should we be concerned? It depends upon which scientist you ask. Some feel that federal standards are too low, and others think that they are too high. The effects of low-level amounts of radiation are quite controversial. Statisticians of high reputation often disagree with each other on this issue. The author has personally witnessed such debates, and the two verbal combatants had the same data in front of them. We are forever asking the question: What is the safe level of radiation? Maybe we need to change the question: Are the risks due to medical applications, occupations, and other low-level exposures insignificant compared to the benefits?

Since we are not certain about whether "safe" radiation really does exist, you might ask why scientists don't simply obtain the necessary data to find the answer once and for all? More precisely, the question being asked is whether the relationship between mutation rate and radiation dose fits the upper graph or the lower graph in figure 9-1. The upper graph indicates that there is no safe dose and that a low probability for mutation exists even with extremely low radiation doses. The lower graph indicates a possible safe dose at very low radiation levels. The low dose parts of both graphs are, for the most part, assumptions, and most of the time they are not based on sufficient concrete experimental data. Scientists tend to disagree about this region of the graph because of problems in trying to collect the appropriate data. Samples must be extraordinarily large or the statistics applied to the data will be meaningless. For example, assume that only one hundred organisms are in the sample, and one mutation occurs at low levels of radiation. This one mutation is possible, but a 1 percent mutation rate is drastically too high. Also, assume that the one hundred organisms in the sample yield no mutations at low levels of radiation. Again this nonevent is possible, but a 0 percent mutation rate is drastically too low. Both results are probable, and both results are wrong. To obtain the necessary resolution, the sample size must be huge. Identifying and working with such large populations is laborious and time consuming. Many scientists, however, feel that no safe level of radiation actually exists, that targets (genes) can be hit regardless of the level of radiation exposure, even at low frequencies. Other scientists actually believe that a little radiation might even be good for you. A consensus among scientists cannot yet be found, so the effect of low-level radiation remains a controversy. Once again, the question of radiation as friend *or* foe arises.

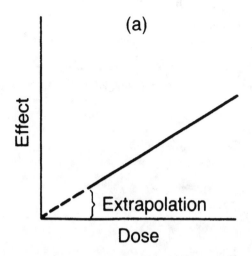

Linear, nonthreshold concept

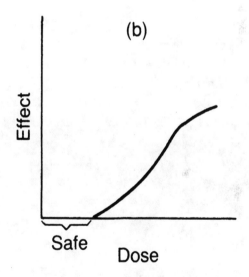

Nonlinear, threshold concept

Figure 9-1. The relationship between the effects of radiation and variable doses of radiation. (a) In a completely linear model, the line is usually extrapolated to zero; (b) in the threshold model, no effects are observed below a certain threshold dose of radiation.

Back in the 1920s, radioeuphoria was ever present. Radium was even added to food and drink. Cocktails would glow in the dim light of nightclubs. People at that time called radiation a metabolic rejuvenation. We have learned much since those days. In fact, the mood today may be better described as radiophobia.

FURTHER READING

Alpen, E. L. 1998. *Radiation Biophysics.* New York: Academic Press.

Calabrese, E. J., ed. 1991. *Biological Effects of Low Level Exposures to Chemicals and Radiation.* Chelsea, MI: Lewis.

Carlson, E. A. 1981. *Genes, Radiation, and Society: The Life Work of H. J. Muller.* Ithaca, NY: Cornell University Press.

Coggle, J. E. 1977. *Biological Effects of Radiation.* London: Wykeham Publications.

Grosch, D. S., and L. E. Hopwood. 1979. *Biological Effects of Radiations.* New York: Academic Press.

Harm, W. 1980. *Biological Effects of Ultraviolet Radiation.* New York: Cambridge University Press.

Kiefer, J. 1990. *Biological Radiation Effects.* New York: Springer-Verlag.

Kondo, S. 1993. *Health Effects of Low Level Radiation.* Osaka, Japan: Kinki University Press.

Lafavore, M. *Radon: The Invisible Threat.* Emmaus, PA: Rodale Press.

Laws, P. W. 1981. *Medical and Dental X-rays: A Consumer's Guide to Avoiding Unnecessary Radiation Exposure.* Washington, DC: Health Research Group.

Schull, W. J. 1995. *Effects of Atomic Radiation.* New York: John Wiley & Sons.

REFERENCES

Committee on the Biological Effects of Ionizing Radiation. 1980. *The Effects on Populations of Exposure to Low Levels of Ionizing Radiation.* Washington, DC: National Academy Press.

Leffell, D. J., and D. E. Brash. 1996. "Sunlight and Skin Cancer." *Scientific American,* July: 52–59.

Muller, H. J. 1927. "Artificial Transmutation of the Gene." *Science* 66: 84–87.

National Academy of Sciences. 1980. *Proceedings of the 15th Annual Meeting of the National Council on Radiation Protection and Measurements.* Washington, DC: National Academy of Sciences.

———. 1981. *Proceedings of the 16th Annual Meeting of the National Council on Radiation Protection and Measurements: Quantitative Risks in Standard Setting.* Washington, DC: National Academy of Sciences.

National Council on Radiation and Measurements. 1987. *Ionizing Radiation Exposure of the Population of the United States.* Report No. 93. Bethesda, MD: NCRP.

Rees, J. L. 2004. "The Genetics of Sun Sensitivity in Humans." *American Journal of Human Genetics* 75: 739–51.

Russell, L. B., and W. L. Russell. 1952. "Radiation Hazards to the Embryo and Fetus." *Radiology* 58: 369–77.

Upton, A. C. 1982. "The Biological Effects of Low-Level Ionizing Radiation." *Scientific American* 246: 41–49.

CHAPTER 10

YOU ARE WHAT YOU EAT— REARRANGED

DNA is messenger-dictating machinery. Transcription is the process whereby your DNA generates these messages. The conglomeration of messages generated is what allows you to digest foodstuffs, carry oxygen in your blood, transform substances into chemical energy, make new cells, repair wear and tear, and generally maintain the life and integrity of all body cells. Genes are the entities responsible for these messages. Genes are segments of the DNA molecule, scattered and sometimes clustered along your DNA. And a DNA molecule can be equated to a chromosome. Some genes are relatively small and other genes can be quite large.

The messages are synthesized on one side or the other of the duplex DNA molecule in the form of RNA molecules. RNA is the acronym for ribonucleic acid, and the molecule is chemically similar to the DNA molecule. The DNA molecule needs to unwind and open up for the generation of the RNA messages. One side of the DNA molecule acts like kind of a template or mold for the RNA message to be built. The transcription process is enzyme mediated, as are most cellular processes. The nature of the message is absolutely dependent upon the sequence of subunits in the DNA segment called a gene. These subunits are four different nucleotides called A, T, G, and C, referring to the nitrogenous bases adenine, thymine, guanine, and cytosine, respectively. If you and I have slightly different DNA sequences, then you and I are turning out slightly different messages. For example, my messages dictate that I should have blue eyes. Your messages might dictate that you should have brown eyes. My messages dictate that I cannot curl the sides of my tongue upward. Your messages might have allowed you to have this great talent. These examples are not a big deal. But some messages are definitely more advantageous than others. Differences are big deals if we are referring to sickle cell anemia, Tay-Sachs disease, ichthyosis, hemophilia, Huntington disease, and so on.

Living a life free of genetic defects is based on the placement of amino acids into a correct sequence. Enzymes run everything in the living organism. If you waited for chemical reactions to occur in your body by mass action, you would not be able to survive. By mass action we mean certain molecules can bump into each other randomly to cause a chemical union, or a molecule can just spontaneously break into two smaller molecules. But accidental occurrences are much too slow to sustain life. We need some way to speed up these reactions. Enzymes do that. Enzymes are catalysts, meaning that they can increase the rate of reactions. That's good because without enzymes life would not exist as we know it. So how do we acquire these enzymes? Everything goes back to the action of genes.

Enzymes are molecules composed of one or more polypeptides. Polypeptides are long chains of amino acids all connected by chemical bonds. Amino acids are molecules that all have certain chemical properties in common but differ from each other by one other property. These differences among amino acids can be as simple as a hydrogen atom or it can be a fairly complex chemical structure.

All living things use the same twenty amino acids in order to build polypeptide chains, one or more polypeptide chains make up a protein, and almost all enzymes are proteins. Not all proteins, however, are enzymes. For the most part, it makes no difference whether you are a mushroom, an earthworm, a water buffalo, or a human. All organisms use the same twenty amino acids to make polypeptides. People are constantly subjected to the names of amino acids because of various health advertisements, health issues in the news, diet supplements, and drugstore shelves. Humans can make some of the amino acids from scratch, but not all of them. The other amino acids must come from breakfast, lunch, dinner, and all of those snacks. The amino acids in the latter group are called the essential amino acids. This description probably means that it is essential for these amino acids to be in your diet. Nevertheless, all twenty amino acids are technically essential whether you make them or have to eat them.

Once again, living organisms need enzymes to conduct chemical reactions and other proteins to carry out specific functions in their cells, tissues, and organs. In order to have a specific function, the twenty different amino acids making up the polypeptide chains must be in a specific sequence. It is the specific sequence of subunits in the messenger RNA that dictates the specific order of amino acids in the polypeptide chain. And the specific sequence of subunits in the DNA dictates the messenger RNA sequence. "Specific" is the key word throughout the synthesis of proteins with the right sequence of amino acids.

Work this series of events in the other direction. Someone with sickle cell anemia, a serious genetic disease, cannot appropriately get oxygen into all of the tissues of his or her body. Of course, that is why the person is anemic. The reason for this oxygen deficiency is due to a distortion of the person's red blood cells. Many of his or her red blood cells are shaped like "sickles" rather than the normal "candy life saver" shape. The distorted red blood cells are more fragile, and they break up and become entangled in the tiny blood vessels called capillaries. The result of these circumstances is impaired blood circulation and decreased oxygen to body tissues. The erroneous shape of the red blood cells, in turn, is due to a defect in their hemoglobin. Hemoglobin is a protein molecule composed of four polypeptide chains, altogether consisting of 574 amino acids. The severe consequence of having sickle cell anemia is due to one amino acid being different at position number six in two of the four polypeptide chains. The amino acid valine is placed into the sequence at that point rather than the amino acid glutamic acid. That one amino acid difference is the result of the wrong RNA message that, in turn, is due to one small mistake (a mutation) in the DNA molecule. Hence, one small change in the DNA can cause a person to be ill equipped to transport oxygen into all of the body tissues. In essence, one small change in the DNA can under certain circumstances be the difference between life and death. Figure 10-1 illustrates a summary of these events.

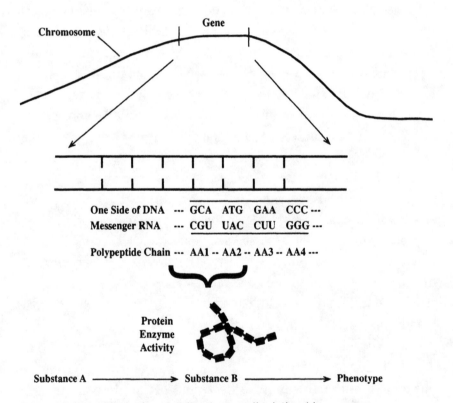

Figure 10-1. An illustration showing the overall relationships among chromosomes, genes, messenger RNA, amino acids, polypeptides, enzymes, metabolic pathways, and phenotypes. Courtesy of Patti Reick.

In a very simplistic way, the rationale underlying the work of our genes can be explained. You eat foods that originate from plants and animals; for example, a slice of roast pork, asparagus, and a baked potato, washed down with a glass of milk. You need to change that pig, potato tuber, asparagus plant, and cow milk into substances that your body can use for energy, maintenance, healing, repair, and growth. First, your body digests those plants and animals that you ate into their component parts, that is, simple sugars, amino acids, fatty acids, and other small molecules, the so-called building blocks. Next, your body needs these building blocks to construct the molecules that you need as a human. This is where the genes come into play. Your genes will direct the synthesis of the enzymes that are needed to place these building blocks into substances that are your body constituents; in other words, the substances become "you."

Most of the time, changing these digested substances, called precursors, into the functional substances you need, called products, requires a number of biochemical steps. Each step requires a different enzyme, and each enzyme requires one or more genes. The biochemical pathway is called intermediary metabolism. In the following hypothetical example, assume that three biochemical steps are required to change the precursor (A) into the product (D).

$$\text{Precursor (A)} \rightarrow \text{Substrate (B)} \rightarrow \text{Substrate (C)} \rightarrow \text{Product (D)}$$
$$(1) \qquad\qquad (2) \qquad\qquad (3)$$

Some metabolic pathways have many more steps before the essential product is finally put together. Note in figure 10-2 that the displayed pathway requires three different enzymes and, therefore, at least three different genes in order to get precursor (A) to undergo the appropriate changes to become product (D). Also note that if you have a mutation in gene *a*, called a metabolic block, you will not be able to make product (D). The consequence of this depends upon how important product (D) is to you. It may be completely inconsequential; or it may cause you to look abnormal in some way; or it may cause you to die at an age much earlier than expected. A mutation in gene *b* results in the same situation; that is, you are still without product (D). Also, it doesn't do any good to have plenty of enzyme *b* from gene *b*, if you do not have any substrate (B) to work on because of the disabled gene *a*. This situation will also be the case for a mutation in gene *c*. Therefore, three different organisms can have a mutation in three different genes, *a*, *b*, and *c*, and all of these organisms will have the same mutant appearance; that is, they will all be without product (D).

The example depicted refers to recessive mutations; therefore, still other genetic consequences are possible. Assume that two persons, *D-less* because of a mutation in gene *a*, mate. All of their children will obviously be *D-less*. A mating between two *D-less* persons due to mutations in gene *b* will also result in *D-less* progeny. The same is true for two *D-less* persons, both of whom have mutations in gene *c*. However, when a *D-less* person due to a mutation in gene *a* mates with a *D-less* person having a mutation in gene *b*, their progeny will all be normal. The person with the bad gene *a* will provide a good gene *b* for the progeny, and the person with the bad gene *b* will provide a good gene *a* for the progeny. All of the metabolic steps will consequently be accomplished. Of course, the progeny will be heterozygous for both gene pairs, but so what. Dominance will allow these progeny to escape being *D-less* (figure 10-3).

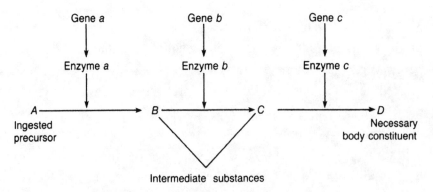

Figure 10-2. A hypothetical example of intermediary metabolism.

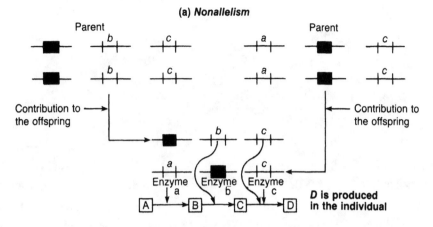

Figure 10-3. Two parents can both be *D-less* as a result of their homozygosity for a recessive allele that blocks an intermediary step in the production of substance *D*. When their metabolic blocks relate to different steps in the pathway, their children will be heterozygous for both metabolic steps and the children will be able to synthesize the product *D*.

A photograph was once published (unknown source) that showed a husky young man of normal stature holding his dwarfed mother in his left hand and at the same time holding his dwarfed father in his right hand. The mother and father were both dwarfs, but because of different gene mutations. Most of the time, albino-with-albino type matings result in albino children. Obviously, this means that most albino people are homozygous recessive for the same gene mutation. On rare occasions, this situation is not necessarily the case. One albino couple was told by their medical doctor not to have children because their children would also be albinos. Albinism causes more difficulties than simply the albino appearance. These individuals can have eye problems, show sensitivity to sunlight, and so on. This particular couple ignored their doctor's advice and had a child, a normally pigmented child. Since that birth went so well, they had a second child, and again the child was normally pigmented. When asked why they went ahead and had two children in spite of all the advice given to them not to have children, they offered the following explanation. They thought that they might be different types of albinos, and they were absolutely correct. They were different albinos. Their mutations were in different genes, and their progeny greatly benefited from this genetic situation. How they could sense that they were different albinos is beyond explanation. Lucky guess?

Phenylalanine is one of the twenty amino acids obtained through your diet. The body normally knows how to convert this amino acid into tyrosine, another amino acid (figure 10-4). If your body cannot effect this conversion, a buildup of phenylalanine occurs, resulting in brain damage and the disorder called phenylketonuria (PKU). This disorder is a metabolic block (block 1 in the diagram). If you have tyrosine, but also have either block 2 or 3, you cannot make melanin, and you will have the albinism disorder. Metabolic block 4 gives you tyrosinosis, and metabolic block 5 means that you are going to have alkaptonuria (black urine disease). These metabolic blocks and their corresponding disorders are due to nonfunctional genes. This is the way metabolism, genes, mutations, and genetic disorders are all connected.

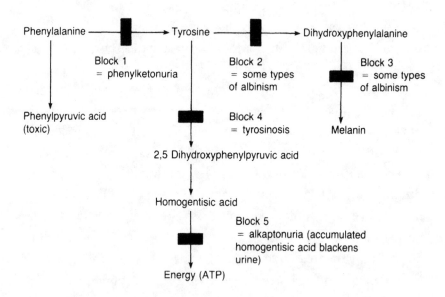

Figure 10-4. Metabolic blocks in the intermediary metabolism of phenylalanine and the serious genetic disorders associated with these blocks.

Humans, monkeys, guinea pigs, and the Indian fruit bat cannot make vitamin C. Most other mammals are genetically capable of making vitamin C from the precursors that they eat. There is no need to feed oranges to a dog. Humans have a metabolic block that disallows the synthesis of vitamin C. We would have scurvy if we didn't eat vitamin C by way of citrus fruits and some vegetables (or vitamin C tablets). Consequently, some geneticists regard scurvy as another genetic disorder that affects absolutely everyone. You may remember as a youngster how you were forced to sit at the table for long periods of time until you ate your broccoli, or some other green stuff. And you might have looked at the husky dog and wondered how the dog grew so big and healthy without broccoli. Fido can make vitamin C. Cats can make vitamin C. Rats can make vitamin C. But you cannot make vitamin C. You must eat your fruits and vegetables because we all carry that metabolic block. Many sailors of early times went without fruits and vegetables for long periods and proved the importance of eating vitamin C, since they suffered the consequences of scurvy.

Living organisms can synthesize large molecules and aggregations of molecules because living organisms own information. That information is in the form of their DNA. If you are old enough, you might remember the popular TV program of many years ago about a bionic man, *The Six Million Dollar Man.* Someone once calculated what a human body was really worth if we completely rendered the body down to a pile of carbon, oxygen, nitrogen, iron, copper, phosphorus, calcium, magnesium, and other elements. At that particular time, the human body was determined to be worth 97 cents on the free market. Someone else decided to make the calculation of the worth of the human body in a different way. This person determined the cost of the large molecules that we synthesize in our bodies like hemoglobin ($2.95 per gram), trypsin ($36.00 per gram), insulin ($47.50 per gram), acetate kinase ($8,860.00 per gram), and so forth. (A gram is about 1/28th of an ounce.) When tallying up the costs of these enzymes, hormones, growth factors, and other expensive molecules, the average-sized body was actually worth an amazing $6,000,015.14. What a coincidence! Everyone is a six-million-dollar person, and with inflation since then, probably much more. The bottom line is that information and syntheses due to this information make us very expensive beings. And that information is all stored in your DNA molecules.

FURTHER READING

Garrod, A. E. 1909. *Inborn Errors of Metabolism.* Oxford: Oxford University Press.

Goodsell, D. S. 2009. *The Machinery of Life.* New York: Springer-Verlag.

Wyngaarden, J. B., and D. S. Fredrickson, eds. 1960. *The Metabolic Basis of Inherited Disease.* New York: McGraw-Hill.

REFERENCES

Chatterjee, I. B. 1973. "Evolution and Biosynthesis of Ascorbic Acid." *Science* 182: 1271.

Garrod, A. E. 1902. "The Incidence of Alkaptonuria: A Study in Chemical Individuality." *Lancet* 2: 1616–20.

Ingram, V. M. 1957. "Gene Mutations in Human Haemoglobin: The Chemical Difference between Normal and Sickle Cell Haemoglobin." *Nature* 180: 326–28.

———. 1958. "How Do Genes Act?" *Scientific American* 198: 68–74.

Pauling, L., H. A. Itano, S. J. Singer, and I. C. Wells. 1949. "Sickle Cell Anemia, a Molecular Disease." *Science* 110: 543–48.

Perutz, M. F. 1964. "The Hemoglobin Molecule." *Scientific American* 211: 64–76.

CHAPTER 11
WOMEN IN THE OLYMPICS

Gender verification for women began in the Winter Olympics of 1968. At first, the tests were carried out randomly, but eventually all women participants were tested. Evidently the objective was to prevent deception, that is, men masquerading as women. The Olympic Committee didn't want anyone with a Y chromosome, the so-called guy chromosome, in the women's competition. Their thinking was that these individuals would have an unfair advantage because of male musculature, increased stamina, and other anatomical differences.

In past years, a few cases of gender deceit did indeed occur. The "female" 100-meter sprint champion in 1932 died in an accident in 1980. The autopsy revealed "her" to have testes. A "female" 400- and 800-meter champion in 1964 was shown to be a male at a later time. A "female" who placed in the women's skiing competition in 1966 later married a woman and became a father—quite an Olympic feat. Another former contestant admitted to passing himself off as a woman under a Nazi order in the pre–World War II years. He was awarded fourth place in the women's high jump event. Suspicions surrounded several others before chromosome testing was invoked. In the mid-1960s, there was much suspicion concerning two "women" of the Soviet Union who were winners of gold medals in track and field. The two champions quickly retired when gender testing began.

Beginning in 1972, buccal cells (inner mouth lining) or root hair cells were analyzed for the presence of a Y chromosome. However, the technique of using buccal cells was deemed to be not very accurate; thus, the root hair procedure became the acceptable test. Each root hair provided about five hundred suitable nuclei for analysis. A special stain was used that makes part of the Y chromosome, if present, emit fluorescence when viewed with a special microscope and light having a certain wavelength. Eventually a different technique called polymerase chain reaction (PCR) was employed in Olympic testing. This is a molecular procedure to analyze DNA, and it could quickly determine whether the SRY gene was present in the individual. Recall that the SRY gene is the sex-determining region on the Y chromosome instrumental in the development of maleness.

About one in five hundred female athletes were dismissed from the competition because they carried the putative male characteristic. The females who were forced to withdraw from the Olympics and go home were, of course, quite humiliated. Their boyfriends were often surprised by the disqualification of their significant other. Females felt that the entire procedure was demeaning and abusive. Very important, none of these individuals were masquerading males. Most of them were very female-like relative to body contour and other female endowments. They simply had a gene or chromosome problem.

The genetics and physiology of gender is very complex. Females with a Y chromosome are not males. For example, some of these individuals had testicular feminization. This genetic problem is a condition whereby XY individuals develop as females. A testicular streak of tissue is generally located within the abdomen, and they also have a blind vagina. The cause of testicular feminization is believed to be a mutation on the X chromosome. These individuals, along with most of the others with gene and chromosome abnormalities, are usually androgen insensitive. Androgen is a male hormone characteristic, and this insensitivity means that most of the individuals did not possess increased muscle strength or any other physical advantage. Regardless of these chromosome glitches, they develop as females. In fact, such chromosome problems usually make them less physically fit than normal females. Consider individuals with Turner syndrome. These persons are XO, that is, without a Y chromosome (so not male) and without a second X chromosome (so not female). Does this problem mean that a Turner syndrome person should be eliminated from the Olympics as neither sex?

Geneticists and many medical personnel criticized gender verification in Olympic competition. Over the years, serious backlash began to creep into the use of these procedures. The whole process was deemed by many to be entirely unnecessary. Finally, in 1999, gender verification was discontinued previous to the Olympic games in Sydney, Australia. Most scientists applauded the change. Those who run the Olympics finally got the genetics right.

FURTHER READING

Leder, J. 1996. *Grace and Glory: A Century of Women in the Olympics.* Chicago: Triumph Books.

Schaffer, K., and S. Smith, eds. 2000. *The Olympics at the Millennium: Power, Politics, and the Games.* New Brunswick, NJ: Rutgers University Press.

CHAPTER 12

MARRYING YOUR COUSIN

Inbreeding represents another important genetic topic having societal implications. Inbreeding is a broad term referring to mating between individuals who are genetically more closely related than when mates are chosen at random. Naturally, the degree of relatedness, called the intensity of inbreeding, can vary immensely. In human genetics, the adjective "consanguineous" describes matings between individuals who have a recent ancestor in common. Generally, consanguineous refers to second- or first-cousin mating. Incestuous activity usually refers to matings between very closely related persons, such as between a parent and child or brother and sister. Most human societies regard such incest as taboo. However, cousin marriages are not infrequent in many parts of the world. In the human population, mating between relatives is of particular concern because of its possible medical and societal consequences.

Many states have laws prohibiting marriages between individuals who are genetically related as first cousins or closer, but a number of other states still allow first-cousin marriages. Legislators in Wisconsin showed some genetic savvy. First cousins are allowed to marry if the woman is fifty-five years old or older. Second-cousin marriages are generally viewed as acceptable in most societies. Various consanguineous mating patterns are illustrated by the pedigrees in figure 12-1. Regardless of the presence of laws prohibiting certain marriages among relatives, states cannot always enforce their laws effectively. Various religious rules and customs are intertwined with the civil regulations in human societies. Some historians believe that the initial laws prohibiting consanguinity were set down for social reasons such as breaking up property ownership and wealth, and not necessarily for genetic reasons.

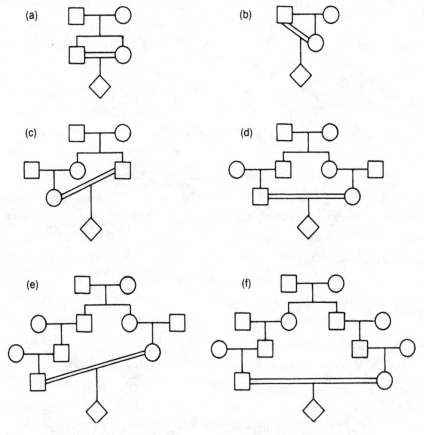

Figure 12-1. Examples of incestuous and consanguineous types of mating: (a) brother-sister; (b) father-daughter; (c) uncle-niece; (d) first cousins; (e) first cousins once removed; and (f) second cousins (second-cousin mating is not deemed a serious genetic matter by most researchers).

The frequency of first-cousin marriages in the United States and Europe is fairly low. Estimates range from 0.5 percent (one out of 200) to 0.1 percent (one out of a 1,000), varying considerably from one region to another. At one time, first-cousin marriages were looked upon with high favor in Japan, and the rate was more than 4 percent. Such social acceptance is no longer the norm; thus, the rate has decreased considerably. Uncle-niece marriages are still favored in some regions of India, where the rates reach more than 10 percent. High levels also exist in some parts of Pakistan. In some regions of Asia and Africa, consanguineous marriages are as high as 20 to 50 percent of all unions. Overall, consanguineous marriages are decreasing because of increased mobility, laws prohibiting such marriages, and changing attitudes. Rates are less than 1 percent in most societies.

Small populations are more subject to inbreeding. Some of the isolated religious communities scattered throughout North America are examples. Because of religious and social customs, their members seldom marry outside their group. Thus they have become genetic isolates by choice. One study of Amish isolates found that a little more than 6 percent of the marriages were between first cousins or closer relatives. When marriages between second cousins were included in the study, the figure became 49 percent. In genetic terms, gene flow seldom takes place outside the isolates. Other genetic isolates have existed simply because of geographic barriers like small island groups of people.

The consequences of inbreeding emerge from the basic concepts of Mendelian and population genetics. The principal genetic effects of inbreeding upon an individual or a population of individuals is an increase in gene homozygosity and a corresponding decrease in gene heterozygosity. The closer the relationship between parents, the more gene forms (alleles) they will have in common, and the greater the tendency is toward homozygosity. By increasing homozygosity, inbreeding fosters the expression of recessive alleles that are normally hidden in the heterozygous condition. Once the alleles are homozygous, the recessive traits are expressed, and some of these traits are very detrimental.

The number of deleterious alleles per individual in a population makes up part of the population's genetic load. Researchers have estimated this genetic load in terms of the average number of detrimental (even lethal) genes per individual. Estimates range from one to four (or even more) serious deleterious genes per person, probably with an average of about two or three per person. Inbreeding increases the chance that such alleles will become homozygous in progeny. No matter how rare a particular recessive allele might be in a population, if a close relative carries the allele, a high probability exists that you might also carry that allele. If the two relatives mate, they face a higher-than-normal probability that their child will express the trait governed by the recessive allele. By similar reasoning, when a trait appears repeatedly in consanguineous marriages, it is usually safe to assume that the trait is caused by recessive alleles following a genetic pattern of inheritance. This relationship is especially obvious for extremely rare traits. Some recessive traits are so rare that the only way the trait can show up in the population is through consanguineous mating. However, the traits need not be recessive. Similar results can even be observed for more complex traits even though their inheritance is complicated.

On a good note, plant and animal breeders use inbreeding to produce inbred strains. The mating of close relatives for many successive generations yields organisms that are homozygous for nearly all of their gene pairs. For example, brother-sister mating for twenty generations will yield organisms in which almost 99 percent of their genes are homozygous and identical among the members of the strain. Developing inbreds in plants is even faster because they can be selfed (sperm to egg within the same plant). When nearly complete inbreds are obtained, the chromosome-segregation pattern at meiosis is always the same in the parents, and the progeny are all genetically identical to each other and to the parents. Numerous plants and animals have been developed in this way for agricultural and other domestic purposes, and some of them are superior strains. Inbred strains have also been developed in mice, rats, rabbits, and other organisms for extensive laboratory research.

On the other hand, however, society tends to be concerned about inbreeding in humans because, as previously explained, the practice increases the likelihood of genetic syndromes. The long-term detrimental effects of inbreeding go beyond specific traits. One study of a large human population over a period of almost a century showed the following results. Offspring from unrelated parents had about a 13 percent mortality rate by age sixteen. Offspring of first-cousin marriages had a mortality rate of 22 percent by age sixteen. Offspring of closer unions had a 32 percent mortality rate by age sixteen. Yet inbreeding by itself, in a strict genetic sense, is *intrinsically* not good or bad. Inbreeding is only a bad thing if the mating individuals have deleterious genes in their genetic makeup. Again, inbreeding simply increases homozygosity, and the effects of the recessive alleles it makes homozygous can be either beneficial or detrimental. More often it is the latter, but there is also a chance that inbreeding will bring beneficial traits to the surface. Such inbreeding success stories are common in domestic plants and animals. History also records examples of consanguinity in humans without many ill effects, as when the Egyptian pharaohs favored brother-sister mating. Yet the odds overall are not favorable, and the successes of plant and animal inbreeding call for careful culling out of rejects. This side of inbreeding is certainly not acceptable to humans. We must care for all of our live-born progeny, live with them, and suffer with them if need be. Once our children are born, we do not eliminate certain ones in the same culling fashion as we do with plant and animal inbred organisms.

First cousins have two grandparents in common. Consequently, the cousins share many alleles derived from these grandparents, and offspring from first-cousin mating can become homozygous for these alleles (figure 12-2). If one member of a first-cousin marriage is definitely a carrier of a particular deleterious allele, the probability that the other first cousin carries the same deleterious allele, identical by descent, is one in eight (1/8). Therefore, the probability that any child of this first-cousin mating is homozygous for the recessive allele in question is one in thirty-two (1/32). This figure is established because the probability that both cousins are heterozygous is 1/8; the probability that two heterozygous parents will have a homozygous recessive child is 1/4; hence, 1/32 is derived from 1/8 multiplied by 1/4.

Data are available that support the theoretical consequences of inbreeding. Children with serious defective traits are often the result of consanguineous mating. For example, children of parents related to each other seem to be more often mentally challenged. Also, data show that related reproducing mates have more miscarriages and stillbirths, and their children suffer more deaths as infants and juveniles. Generally, the risk of having children with a serious genetic

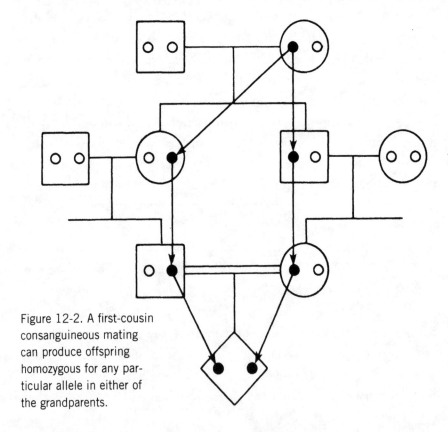

Figure 12-2. A first-cousin consanguineous mating can produce offspring homozygous for any particular allele in either of the grandparents.

defect to unrelated parents is 3 to 4 percent. Considering the same types of defects, first-cousin matings have shown the risk to increase to about 6 percent. The percentages vary slightly from study to study, but the risks for first cousins are almost twice as high as matings between nonrelatives. Some scientists do not consider the difference between first-cousin and nonrelative matings to be very significant, perhaps because they see the difference as being only 2 or 3 percent. Therefore, some actually feel that the risk is not high enough to discourage such marriages and that the law, where it exists, should be abolished. Nonetheless, the risk is almost a twofold increase. Lowering the frequency of consanguineous mating would surely reduce the frequency of genetic defects to some extent and improve the survival rates of children. The difference is even more pronounced for major genetic abnormalities in matings of closer-related individuals. Although marriages between individuals closer than first cousins are not legal in North America and the data are limited, such matings do occasionally occur. Uncle-niece and aunt-nephew matings result in about a three- to fourfold increase in birth defects, and even higher (almost 40 percent) increases are observed for parent-child and sib-sib matings.

The following is a hypothetical case. Robert and Richard Brown are monozygotic (identical) twins. Susan and Sally Green are also monozygotic twins. Robert marries Susan and Richard marries Sally. A son, Roland, was born to Robert and Susan, and a daughter, Sarah, was born to Richard and Sally. All of these happenings took place in a state that allows first-cousin marriages, and Roland and Sarah wish to marry each other. Roland and Sarah, although socially are first cousins, they are genetically brother and sister. This marriage should not happen. A 40 percent risk of having progeny with serious genetic defects would be facing them. Or consider this situation. One of your best friends has a problem and comes to you for advice. His son wants to marry the girl next door. Your friend, however, happens to also be the father of the girl next door (unknown to everyone except the mother of the girl). Now is the time for the wayward one to come forward and be brutally honest. The boy and the girl next door face a significant risk of their children having a genetic defect because they could have deleterious recessive genes in common with each other.

REFERENCES

Bodmer, W. F., and L. L. Cavalli-Sforza. 1976. *Genetics, Evolution, and Man.* San Francisco: W. H. Freeman, 364–77.

Jenkins, J. B. 1990. *Human Genetics.* New York: Harper & Row, 462–67.

Mange, A. P., and E. Johansen Mange. 1980. *Genetics: Human Aspects.* Philadelphia: Saunders College, 461–83.

McConkey, E. H. 1993. *Human Genetics: The Molecular Revolution.* Boston: Jones and Bartlett, 131–33.

Seemanova, E. 1971. "A Study of Children of Incestuous Matings." *Human Heredity* 21: 108–28.

Stine, G. J. 1989. *The New Human Genetics.* Dubuque, IA: Wm. C. Brown, 420–22.

CHAPTER 13

THE BABY IS NOT MINE

Disputes over biological relationships often end up in court. Occasionally, relatedness must be established to settle a will. Although the event is uncommon, when mothers are given the wrong babies in the maternity ward, the legal system must have some way of determining which baby belongs to which mother. The same kind of question arises more often in divorce and child support cases. These cases have to be decided according to whether the putative parent, usually the father, is actually the genetic parent of the child.

British researchers have carefully analyzed decades of genetic test results of people in the Western world, and they have found paternal discrepancies galore. They have established that approximately 4 percent of all cases studied showed potential parentage discrepancies. Thus, about one in twenty-five people had a biological father other than the one listed on the birth certificate. One in twenty-five! Sometimes medical researchers come across genetic discrepancies but leave the situation unmentioned in order to avert family problems. Increasingly, however, men are requesting genetic tests to prove or disprove fatherhood, especially in paternity suits.

Previously, and over many years, blood group data were used as evidence in courts in order to possibly resolve paternity questions. However, these tests showed evidence that only excluded paternity. The tests did not indicate definitive paternity; that is, blood tests only determined that a person was not the father, but these tests could not show that the person absolutely was the father.

Blood groups as tests of paternity are nonetheless based on sound Mendelian genetics. For example, the ABO blood group system was one of several different blood group systems used for this purpose. This blood group is a multiple-allelic system controlled by three different alleles (forms of a gene), called A, B, and O. Blood groups A and B are dominant to the O blood group. Blood groups A and B are codominant relative to each other. Codominance means that the alleles together yield an AB genotype. Both antigen products, A and B, can be detected in the red blood cells of the individual

TABLE 13-1

Possible and impossible ABO phenotypes of offspring resulting from all possible matings with respect to ABO blood groups

Phenotypes of Matings	Possible Offspring	Impossible Offspring
A × A	A, O	B, AB
A × B	A, B, AB, O	none
A × AB	A, B, AB	O
A × O	A, O	B, AB
B × B	B, O	A, AB
B × AB	B, A, AB	O
B × O	B, O	A, AB
AB × AB	A, B, AB	O
AB × O	A, B	AB, O
O × O	O	A, B, AB

because there is no dominance involved. These dominance and codominance relationships determine the offspring that a particular mating can and cannot produce, as shown in table 13-1.

Still other blood groups are known. Some of them, such as the MN system, were also used in assessing parentage. M and N is another case of codominant alleles, and the system is made up of MM, MN, and NN genotypes and corresponding M, MN, and N phenotypes. The Rh blood group is complex, but it was still a medical-legal useful system. Everyone can be classified as Rh+ or Rh−, with Rh+ being dominant to Rh−. The numerous possible combinations of different blood groups added to their usefulness in settling parental disputes.

Consider the following example. A woman had Rh− and O blood types. The putative father had Rh+ and AB blood types. The infant was Rh+ and A. In this case, the man could not be excluded as the father of the child because genetically he could possibly be the father. However, approximately 38 percent of all males in the United States have blood group type Rh+ along with either A or AB blood type. So once again, blood types are not definitive as to who really is the father. In one confirmed case, a woman had fraternal twins, and the blood tests showed that the twin members had two different fathers. This woman obviously got around. In past years, rich men were often sued by poorer women for child support because parental testing was not definitive. Even the famous Charlie Chaplin was named in such a case, although blood types indicated he was not the father.

Other genetic markers were ultimately used for paternity testing. Probabilities reached the "practically proven" level of 99.8 to 99.9 percent by including these other genetic markers. One complex group of such markers comprises the antigens of white blood cells, including the human leukocyte-associated (HLA) antigen system. Many different white blood cell genetic markers are known. Because each combination of HLA alleles is rather unique, a match between a putative father and a child was taken as strong evidence for parentage.

Most courts eventually accepted such genetic information for assigning parentage. But consider this interesting situation. A woman with type O blood had a type A husband, and she gave birth to a child with type B blood. One might immediately be prone to denounce this woman as being unfaithful. Or one could suppose the discrepancy to be due to a mutation of the O allele in which the allele changed to a B allele in the woman. Such a change, however, is highly unlikely because mutations are extremely rare. Checking the results several times also ruled out human error. Another genetic situation, although quite rare, actually explained this discrepancy. People synthesize their A and B substances (antigens) in the blood by converting another substance, called H, to A or B. Therefore, humans require another enzyme in order to make the H substance that is the precursor for the A and B antigens. If an individual is deficient for this particular enzyme, there will be no H substance to convert to A and B, regardless of having good genes for A and B. Hence, the individual will test as an O blood type. Figure 13-1 illustrates this pathway and the blood type results of the various gene combinations.

The woman with the O blood type had nonfunctional genes for H and A and, consequently, could not make B either, although she had a functional gene for B. However, she could pass this functional gene for B to her child. The cross between these two parents involves the following phenotypes and genotypes.

Mother (blood type O)   Father (blood type A)
hh and (B) O   HH or Hh and AO
Child (blood type B)
Hh and BO

Persons lacking the alleles to make the H substance are described as having the Bombay blood type. At any rate, this couple was able to avert domestic strife. Good ending!

Today we have much better techniques to examine each individual's unique genetic composition—that is, by profiling DNA. This

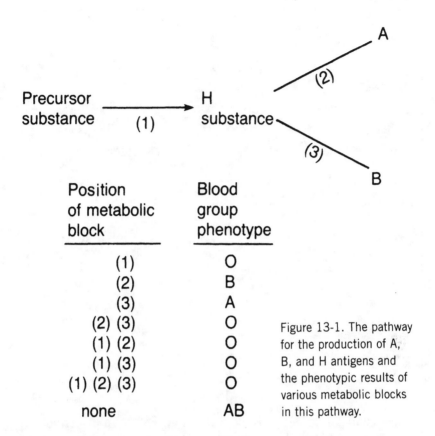

Figure 13-1. The pathway for the production of A, B, and H antigens and the phenotypic results of various metabolic blocks in this pathway.

Position of metabolic block	Blood group phenotype
(1)	O
(2)	B
(3)	A
(2) (3)	O
(1) (2)	O
(1) (3)	O
(1) (2) (3)	O
none	AB

technique, called DNA fingerprinting, can give nearly 100 percent definitive identification. Although DNA fingerprinting has many applications, one of the most common usages in the United States has to do with establishing correct paternity situations. Probably several hundred thousand paternity questions are resolved by DNA finger-printing each year, many with regard to births by single mothers.

An English scientist, Alec Jeffreys, first developed DNA finger-printing. The technique is based on the knowledge that we all have hypervariable regions in our DNA. These regions of DNA, often called repetitive DNA, differ greatly among individuals. Recall that DNA is a sequence of subunits simply referred to as A, T, C, and G. The specific pattern, length, and number of these repetitive sequences among individuals are remarkably different in these

TABLE 13-2

Hypothetical data of four DNA regions

Hypervariable DNA region	Frequency of the DNA pattern in the population	Probability when the DNA regions are considered together
1	1/250	_____
2	1/100	$1/250 \times 1/100 = 1/25,000$
3	1/320	$1/250 \times 1/100 \times 1/320 = 1/8,000,000$
4	1/90	$1/250 \times 1/100 \times 1/320 \times 1/90 = 720,000,000$

hypervariable regions. Jeffreys developed techniques for analyzing the DNA in these hypervariable regions. When several of these DNA sequences from an individual are analyzed, the patterns taken together become extremely unique relative to other individuals (table 13-2).

Since the work by Jeffreys, several other techniques have been developed to conduct DNA fingerprinting. Most methods, however, are based upon the hypervariable regions that people have scattered throughout their repertoire of DNA. These sequences are so variable that it would be extremely rare that two nonrelated individuals would have an identical set of sequences. Of course, identical twins and other identical multiple births are exceptions since they theoretically have identical DNA.

The child receives one set of DNA patterns from the mother and one set from the actual father. This situation does not diminish the validity of the test. The child will share half of his or her DNA patterns with the father, and that can be easily analyzed. The resultant patterns are much like the bar codes found on retail items (figure 13-2). With appropriate standards and controls, DNA fingerprinting has become an absolute parental identification, rather than just parental exclusion, as shown with other genetic tests.

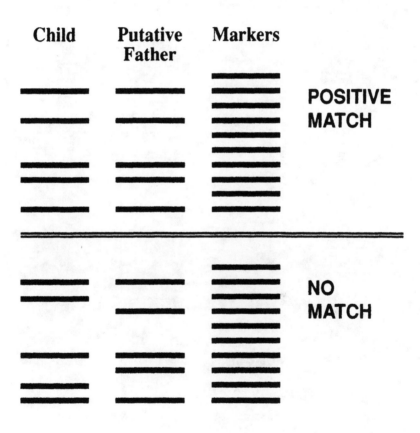

Figure 13-2. A hypothetical view of DNA fingerprinting results. The column on the right indicates the sizes of the DNA pieces being analyzed. Note that a perfect match between the child and the putative father exists in one case but not in the other case. Courtesy of Patti Reick.

Figure 13-3. Family Circus © Bil Keane, Inc.
King Features Syndicate. Reprinted with permission.

REFERENCES

Blende, Y. M., C. K. Deshpande, H. M. Bhatia, R. Sanger, R. R. Race, W. T. Morgan, and W. M. Watkins. 1952. "A New Blood Group Character Related to the ABO System." *Lancet* 1: 903–904.

Sussman, L. N. 1973. "Blood Grouping Tests for Non-Paternity." *Journal of Forensic Sciences* 18: 287–89.

Wiener, A. S., and W. W. Socha. 1976. "Methods Available for Solving Medico-Legal Problems of Disputed Parentage." *Journal of Forensic Science* 21: 42–64.

CHAPTER 14

LOSING THEIR PUNCH

Antibiotics are losing their punch as bacterial populations are changing to resist them. For example, penicillin was extremely effective against many different bacteria at one time. Today, the antibiotic is only effective against a small number of bacteria. And the more we use these antibiotics, the more we increase bacterial resistance to them. We are, in a sense, creating "super bugs." The same story is being played out with other drugs used against diseases, insecticides against insect pests, herbicides against noxious weeds, and crop plants bred against plant pathogens. The explanation for this loss of punch is mutation, leading to genetic resistance, leading to the survival of the bugs no matter what we throw at them.

Mutation is the ultimate source of all diversity. Selection, on the other hand, is probably the greatest driving force effecting changes in a population. Selection can cause the proportions of gene frequencies to change in a population. Selection is categorized as two types: natural and artificial. Both types of selection can be an effective means of bringing about gene changes in a population of organisms. Natural selection is the process whereby nature decides which organisms will survive long enough to reproduce, that is, which organisms will leave their genes behind to the next generation. With artificial selection, humans make the choices relative to which organisms will be allowed to produce the next generation. Artificial selection is especially powerful in creating changes because its purpose is to concentrate on certain traits. Artificial selection is the time-tested and proven method of plant and animal breeders.

A tremendous amount of genetic variability exists in organisms. This variability comes about because many of the genes possessed by an organism can have many different forms, again called alleles. Such inherent variability is the basis for the effectiveness of artificial selection. For example, consider Morrow Fields located on the campus of the University of Illinois. These fields have been the site of classic experiments going on since 1896. In one case, plant breeders selected for high protein content in the kernels of corn. In another case, they selected for high oil content in the kernels. In both cases, the levels consistently increased, not counting some fluctuation from year to year due to environmental conditions. The resultant corn is not particularly good relative to yield, harvest, and storage characteristics, but the results demonstrate a genetic point. Genetic variability within organisms is immense, and therefore selection can indeed be powerful.

Next, consider the case of breeding mice in order to increase their size. This genetic researcher kept crossing the largest mice with each other for twelve years (about 40 generations) until they were over twice as large as normal mice. Someone asked this investigator how large he expected the mice to ultimately get, and he responded, "How do you think elephants got so large?" Eventually one comes to realize the point he was making, which was that selection leads to evolution. Another researcher bred *Drosophila* (fruit flies) to become completely oblivious to DDT, one of the most effective insecticides ever produced (now banned from general use). The flies actually played in solid DDT powder. Another fruit fly researcher needed to get a large population of the little critters from a large breeding box where they underwent massive proliferation into a bottle for carrying out further steps in his research protocol. The use of light and other inducements to coax them into the bottle just did not work very well. He then remembered that he was a geneticist, so he bred fruit flies that loved to go through holes. When he opened a hole in the breeding box, the flies literally rushed into his container without any coaxing. Because of their short life cycle (about twelve to fourteen days), it only took him about six months to breed a population of flies that were obsessed with flying through holes.

Natural selection is slower than artificial selection, but nonetheless it is an effective mechanism for fueling evolution. Evolution simply means that populations of organisms change over long periods of time. The concept was set down in 1859 by the famous Charles Darwin (*On the Origin of Species by Means of Natural Selection, or the Preservation of Favored Races in the Struggle for Life*). Alfred Wallace, a contemporary naturalist of Darwin, concurrently developed the idea of natural selection. In simple terms, the argument for natural selection is set down as follows: (1) organisms tend to reproduce in excess relative to their environmental opportunities; (2) consequently, a struggle for existence ensues; (3) the organisms that tend to survive their environment to mature and reproduce are those best adapted to survive; and (4) if these adaptive traits for survival are hereditary, they will be passed on to their offspring—all of which can lead to changes in the population. The first printing of Darwin's book was sold out by noon the first day. Following Darwin's publication, one wonders how many scientists of that period of time murmured to themselves "Why didn't I think of that?" Many scientists see the concept of natural selection as being both simple and profound at the same time.

The following scenario is a simplistic demonstration of how natural selection works. Assume that a large animal barn is teeming with flies. The flies love the place because of all of those little goodies lying around on the floor. Also assume that someone fumigates this fly population in the barn with an insecticide by heavily pumping the anti-fly stuff throughout the entire structure. This maneuver will probably kill *most* of the flies. However, a few flies may escape this attempted purge, and most important, because of genetic reasons. A few may be able to close their spiracles (little breathing tubes on the sides of their abdomen) for a long enough period of time in order to safely get out of the barn. A few may have an uncanny ability to hide in small cracks and fissures away from the insecticide. Some may even have a different biochemical metabolism that breaks the insecticide down in the body into nonlethal substances. If any of these characteristics are genetic, they can be passed on to progeny. The dead flies, of course, will no longer reproduce, but the resistant flies will; hence, the population will slowly change into one consisting of mostly resistant-type flies.

The development of resistance to adverse environments among organisms by way of natural selection is indeed a continual problem for humans. Not only do we make insects resistant to insecticides, and weeds resistant to herbicides, but also nasty bacteria resistant to antibiotics. In other words, this same scenario is being observed in the world of medicine (figure 14-1). Many bacterial strains are now resistant to many of our antibiotics simply because we overused these wonderful drugs. Medical researchers and drug companies are constantly trying to stay ahead of pathogenic microbes. Agricultural researchers are desperately trying to stay ahead of insects, weeds, and plant diseases. Other researchers are trying to stay ahead of the proliferation of other pests like mosquitoes, fungi, tree-eating insects, and so on. Needless to say, this situation is a war, and it is all thanks to natural selection and genetic resistance.

The idea of survival of the fittest sounds like reasonable events occurring in nature, but like all scientific ideas, natural selection requires evidence. Some of the first evidence supporting the concept of natural selection came from classic experiments and data collected by English scientist H. B. D. Kettlewell. This scientist researched a moth named *Biston betularia*. Two strains of this moth have contrasting color characteristics. One strain, called *carbonaria*, is

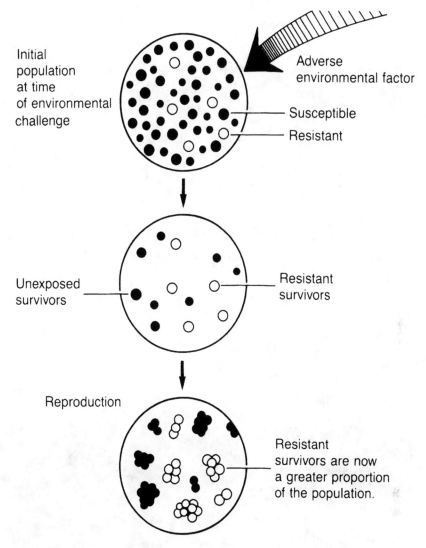

Initial population at time of environmental challenge

Adverse environmental factor

Susceptible

Resistant

Unexposed survivors

Resistant survivors

Reproduction

Resistant survivors are now a greater proportion of the population.

Figure 14-1. A diagrammatic explanation of natural selection as the cause of change in the genetic composition of a population.

dark and the other strain, *typical,* is light colored. The light and dark colors are heritable characteristics. In about 1800, the tree trunks in the wooded area surrounding certain industrial centers of England were mostly light colored because of lichens that covered them. At that time, most of the moths hanging out in those woods were also light colored, blending in very well with the lichens; only a few dark-colored moths could be found. As the Industrial Revolution progressed, the tree trunks became dark colored because of smoke and soot, and then concurrently the frequency of the dark moths in the population greatly increased from rare to about 90 percent. In other words, the gene pool changed because the alleles for light-colored moths became rare, and the alleles for dark-colored moths became common (figure 14-2).

Kettlewell and others showed that the changes in the moth population were due to environmental changes and predatory birds. The color of the moths played a camouflaging and protective role in their survival. Moth-eating birds made lunch out of the light-colored moths against dark tree trunks. The light-colored moths were thus selected out in the sooty environment. On the other hand, with increased environmental concerns in more recent decades, air pollution decreased in these areas of England. Hence, the tree trunks

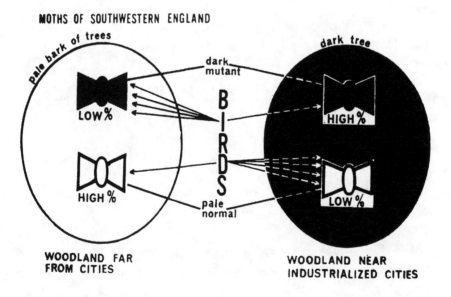

Figure 14-2. A diagrammatic explanation of natural selection playing a role in the frequency of the strains of moths in different wooded regions.

were returned to a light color again, and the birds were now gulping down the dark moths at an increased rate. These changes have been thoroughly studied, even to the extent of filming the actual predatory events taking place by the birds and counting the lunch victims. The moth-bird interaction is now a classic example of natural selection at work.

A very important point must be emphasized with regard to natural selection. The individual moths *did not* change from light to dark because the tree trunks had become dark. Rather, the frequencies of the light and dark moths in the population changed over a number of generations due to natural selection—in this case, the type of selection identified as predatory birds. The individual moths were either adapted or they were not adapted even before any change occurred to the tree trunks. This concept is known as preadaptation, and it is in sharp contrast to postadaptation. Postadaptation means that an individual organism could change when the environment becomes adverse. An organism survives or gets selected out of the population based upon the gene repertoire that it already owns, which is preadaptation. Some flies, for example, were already adapted to survive the insecticide before they were sprayed. Proving preadaptation was not easy. Once the organisms are subjected to an adverse environment, how does one know whether the survivors were adapted beforehand, or all of a sudden they underwent adaptation because of the environmental shift? Nonetheless, several lines of evidence for preadaptation have been reported owing to some ingenious experiments.

Since natural selection is a population change, it obviously doesn't work quickly. One would think that pheasants would eventually quit flying up when hunted; they get shot in their behinds when they fly up from their hiding spot. One would think that dogs would quit chasing cars; they get run over when they chase cars. One would think that mousetraps would eventually cease to work so well; however, these traps seem to be as effective as ever. Mice continue to like cheese, and they continue to get clobbered for it. When thinking about natural selection, human examples can also come to mind like riding snowmobiles on thin lake ice; or getting close to wild buffalo in a national park in order to photograph them; or speed boating without wearing a life jacket; or standing under a lone tree in a thunderstorm; or swimming in shark-infested waters; or rock climbing after a few drinks. These examples may be a little facetious, but one can get the point. First, I doubt that genes exist for any of these traits (except maybe for stupidity). But if these traits were hereditary, the sorry event often occurs too late because the victims may already have had children.

FURTHER READING

Darwin, C. 1859. *On the Origin of Species by Means of Natural Selection; or, The Preservation of Favored Races in the Struggle for Life.* London: John Murray.

REFERENCES

Kettlewell, H. B. D. 1955. "Selection Experiments on Industrial Melanism in Lepidoptera." *Heredity* 9: 323–42.

———. 1956. "Further Selection Experiments on Industrial Melanin in the Lepidoptera." *Heredity* 10: 287–301.

———. 1961. "The Phenomenon of Industrial Melanism in the Lepidoptera." *Annual Review of Entomology* 6: 245–62.

CHAPTER 15

GENETICS IS BIG
IN AGRICULTURE

Genetics has contributed greatly to the increase in crop yields that have occurred over the years. The magnitude of its contribution, however, is hard to estimate. Other factors have also been involved in these increases. Farm management, pest control, fertilizers, and improved farming techniques may be responsible for half of all the increases in yield since the late 1930s, with the other half due to genetics. Geneticists alone cannot feed the world. This endless task requires the efforts of plant and animal breeders, agronomists, and many other scientists as well.

Crop plants were first domesticated and bred by prehistoric people who selected plants with the most desirable traits from the population of wild plants. Today's plant breeders are also constantly seeking improved plant varieties. With major crop plants, breeders usually need an average of about eight to ten years to produce a new and satisfactory variety by traditional plant-breeding methods.

Plant breeders exploit the reproductive and other biological characteristics of plant species. In one basic technique, they cross different strains of the species they are attempting to improve. They then produce subsequent generations of progeny by self-pollination (fertilization) to create new combinations of genes. Using these further generations produced by self-fertilization, they then stabilize the many resultant lines among the progeny so that each line will breed true. Finally, the breeders select the progeny lines whose gene combinations they judge to be best for one reason or another and field-test them, usually in several different environments or locations in the country or even the world. Favorable test results lead to the release of the new variety to farmers.

Many breeders believe that even better results can be achieved. Some think we should make concerted efforts to select plants that are better adapted to environments that are considered unsuitable for present agricultural uses. Such schemes could be very successful because artificial selection can be a powerful influence on the development of organisms.

In 1970, plant breeder Norman E. Borlaug won the Nobel Peace Prize for his role in initiating a "Green Revolution." He was instrumental in breeding high-yield wheat strains for developing countries. These strains have shorter and stronger stems and are much better adapted to the environments of the countries for which they were bred. Given plentiful fertilizer, these plants produce dramatically increased wheat yields. Using these varieties, Mexico changed from being an importer of wheat to a large wheat exporter. India increased its wheat yield an unbelievable 700 percent. Other countries have also used the semi-dwarf wheat variety with success. These new varieties represent a tremendous agricultural achievement and a great contribution to world food supplies. Because food is so essential for both economic growth and political stability, these varieties were deemed helpful for maintaining world peace.

Hybrids are interesting genetic happenings. The connotation of *hybrid* takes on several different meanings, dependent upon what biological discipline is providing the definition. Geneticists and molecular biologists usually think that being heterozygous at any gene pair constitutes a hybrid—for example, Aa rather than AA or aa.

Plant and animal breeders generally think of crossing different strains of the same species to produce hybrids—for example, a cross between two different inbred strains of corn to produce hybrid progeny. Evolutionists and ecologists often think of a hybrid as a cross between two different species—for example, a walleye pike crossed with a sand pike.

In 1800, about one billion people inhabited the earth. By the year 2000, the number of people on earth had increased to over six billion. Imagine the amount of food devoured each day by six billion people. Some researchers feel that we would have starved long ago if we had not significantly increased food production. Much of this increase in food production is due to genetics, and much of the genetics in question is due simply to the development of hybrids. For example, corn production before hybrids was generally about twenty to twenty-five bushels per acre. Today with the use of hybrid corn and newer modes of management, it is easy to obtain one hundred to two hundred bushels per acre (and sometimes even more). Similar situations can be seen throughout agriculture. What a difference it often makes to cross inbreds to obtain hybrids.

From a genetic standpoint, the exact reason for the augmented yield or the greater size of hybrid progeny is still a good argument. Some scientists think that the genetic reason is just plain dominance. For example,

Inbred AA bb CC dd ($\times$) Inbred aa BB cc DD $\rightarrow$ F_1 hybrid Aa Bb Cc Dd

The hybrid results in a dominant allele at each locus, making a better organism than the two inbreds, both of which are homozygous recessive for several gene pairs. On the other hand, some scientists feel that being heterozygous like Aa is actually better than aa and even AA. This circumstance, if real, is called overdominance. Still other scientists think that both dominance and overdominance are involved in hybrid superiority. The greater vigor in terms of growth, yield, survival, and fertility of hybrids is known as heterosis. And the phenomenon of heterosis always seems to be associated with the heterozygosity of genes.

Another factor involved in producing hybrids is combining ability. Bringing the chromosomes together from two different organisms can result in very unpredictable outcomes. One never really knows the outcome in advance of actually doing the cross. It is remindful of the researcher who was able to cross radish plants with cabbage plants. The intended outcome was to obtain progeny that would have a root like the radish and leaves like the cabbage. However, the actual result was a plant with a root like the cabbage and leaves like the radish. Hopes were dashed. Hybrids brought about by crosses between different strains of the same species usually do not have such surprises, but again one cannot always predict the result. Sometimes the hybrid progeny will appear super, and sometimes the hybrid will not appear very good at all. Breeders of plants and animals describe this situation as the "combining ability" between the organisms being crossed. Breeders are always seeking pairs of inbreds that have great combining ability because, of course, they want to produce great hybrids.

Many traits of organisms are due to multiple gene pairs. This mode of inheritance is called polygenic (many genes). Polygenic inheritance is in contrast to traits due to a single gene pair, referred to as Mendelian inheritance. Usually, but not always, polygenic traits are measurable; that is, such traits form a continuous distribution of expression. For example, weight, height, skin color, and IQ in humans are probably polygenic traits. Also, the environment plays a significant role in such traits in addition to genes. This environmental role can be easily seen in the human traits we have listed: weight, height, skin color, and IQ. On the other hand, Mendelian traits are all-or-none situations, meaning that you are either albino or not, showing phenylketonuria or not, having hemophilia (bleeder's disease) or not.

The genes involved in polygenic inheritance, called polygenes, have additive effects on the trait. Thus, the more alleles that exist for a particular trait, the greater will be the expression of the trait, and vice versa. Dominance and recessive descriptions are not involved in polygenic traits. Therefore, some organism with ten alleles for enhanced height would be taller than some other organism with six alleles for enhanced height, not counting the effects of the environment, which in some cases can be a considerable factor in the magnitude of the trait.

Experiments can be conducted with plants and animals to determine the extent to which heredity and environment play a role in a polygenic trait. Researchers accomplish this calculation by growing different genotypes (gene compositions) of the organism in different environments, measuring the trait, and doing the appropriate statistics. The end result is a calculation of the percentage of the variation observed in a particular measurable trait that is due to heredity. This calculation is called heritability. For example, the heritability of back fat thickness on pigs is 70 percent. The pigs' diet and how they are cared for, which make up the environmental component, is the other 30 percent of the variation noted in back fat thickness.

You might recall a TV commercial concerning an ice cream product that concluded with the statement, "It's the cows." Whether the vendor knew it or not, it appears that the statement was referring to the concept of heritability. Plant and animal breeders pay close attention to heritability. If the heritability of a trait is high, the trait can be successfully increased through selection procedures and breeding. If not, the trait is due more to the environment and very much out of the control of the breeders.

Despite their past successes, plant breeders cannot be complacent. Pathogens and insects are constantly evolving, and as a result, they are attacking crops in new ways. Consequently, plant breeders must continuously develop new varieties resistant to natural threats. The breeders are involved in two races: one against the natural enemies of crop plants, and another against world food shortage. In addition to resistance and yield, breeders also seek to improve nutritive quality.

One other component contributing to polygenic traits should be mentioned, that is, gene-environment interaction. Gene-environment interaction is often another factor involved in the expression of a trait, and it is a separate component from genes and the environment considered alone. This interaction is another important component in how the trait might turn out. Farmers certainly know gene-environment interaction very well. They plant certain varieties in their fields dependent upon where they do their farming. Some varieties may do very well in Montana, but not in Minnesota, and vice versa. If farmers didn't pay close attention to gene-environment interaction, they could lose the farm—literally.

Genetic engineering has excited not only medical and pharmaceutical researchers but also agricultural scientists. We even hear predictions of another Green Revolution. If genetic engineering techniques can effectively increase food production and improve crop quality, the impact of these techniques may be greater in agriculture than in medicine. Relatively few people have incurable diseases or require expensive drugs. But everyone eats three times a day and some more often.

FURTHER READING

Bickel, L. 1974. *Facing Starvation: Norman Borlaug and the Fight against Hunger.* New York: Reader's Digest Press.

Ronald, P. C., and R. W. Adamchak. 2010. *Tomorrow's Table: Organic Farming, Genetics, and the Future of Food.* New York: Oxford University Press.

REFERENCE

Borlaug, N. E. 1983. "Contributions of Conventional Plant Breeding to Food Production." *Science* 219: 689–93.

CHAPTER 16

AMAZING MAIZE

The scientific name of maize (or corn) is *Zea mays*. Maize is one of the most efficient plants among all cereal grains in using solar energy to produce food. The species undergoes a biochemistry different from most other plants that makes it efficient, especially in hot, dry weather conditions. The origin of maize is believed to be Mexico or Central America, and it has always been part of the culture of America. In addition to the great agronomic importance of maize, the plant has been used in many nonfood products. Also, the species has been extremely well suited for genetic research ever since the rediscovery of Mendel's classic paper in 1900. Since then a tremendous amount of interest has been given to the species by agronomists, botanists, geneticists, molecular biologists, and plant physiologists, among other scientists.

Maize is the world's second-largest crop plant behind rice. Maize provides 19 percent of the world's calories and 15 percent of the world's protein. Maize as a food is the staple for at least 200 million people. The United States produces over 330 million tons of corn kernels every year, amounting to about 42 percent of the total amount produced worldwide. Much of the maize crop is used for human consumption. And a lot of it is also used for animal consumption, which technically also ends up in human stomachs.

When one thinks of maize as a human foodstuff, sweet corn most certainly comes to mind. Sweet corn is simply a genetic version of field corn. This sweet variety of corn is due to a mutation of a gene that geneticists call sugary. When picked at the right time, the kernel will have that sweet taste everyone is familiar with. However, if one waits too long, the sugar changes to starch, which is another taste everyone has probably experienced. The corn that we eat is high in both carbohydrates and protein.

Most people are surprised at how much corn they really eat beyond the ears of sweet corn, corn chips, and the corn in cans or the freezer. Corn is an ingredient in a vast number of other foods on the grocery store shelves. One study found corn used—either in the foods or in their processing—in 2,500 out of every 10,000 items in a typical grocery store. That's approximately 25 percent of every grocery store

Figure 16-1. An array of some of the common foodstuffs made from corn.

aisle that the customer walks through. Even smut, the ugly black fungus that grows on corn plants in the field, is used as a food in some places like Mexico. As a child, I was told that if you touched the smut it would grow on you. I won't eat it.

Corn production was part of the war effort during World War II. Of course, it provided a great amount of food for the country. This period was about the time that hybrids came on the scene and yields increased dramatically. And corn also contained oil for various wartime uses. After the war, the vast stores of corn helped to save Europe from starvation. This rescue effort was the result of the Marshall Plan. Some historians go so far as to assert that corn was actually a big part of winning the war. Too strong a statement?

Ethanol production is another facet of the use of corn. Ask any moonshiner. Industrial production of ethanol from corn has been steadily increasing. A small amount of ethanol is now added to gasoline. However, a significant amount of concern has been cast with regard to the efficiency of using corn for ethanol production. Other plants are deemed to be more efficient for this purpose. But it "still" works well for moonshine.

Corn can be used as a fuel in another way. It can actually be burned directly in specialized stoves built and adapted for wood and corn. The kernels contain oil, and they burn quite hot without popping and crackling. Since corn, even good corn, is relatively inexpensive, the fuel is cheaper than other fuels, including wood. Corn is actually estimated to be about 2.5 times cheaper than wood. The use of corn-burning stoves is increasing.

Numerous products use corn in some capacity. In addition to foodstuffs, over eighty other products have been listed, such as packing materials, soil stabilizers, adhesives, cosmetics, paints, toothpaste, fabrics, and pharmaceuticals. And don't forget the throwaway beer glasses made of degradable plastic. Some companies are also mining the corn kernel for vitamins, minerals, and components of medicines. Unknown cures may still be lurking in the genetics of those kernels.

Figure 16-2. Many substances have been found
in corn kernels besides carbohydrates.

In addition to all else, maize has been a model organism for the studies of genetics, molecular biology, and developmental biology. Many genetic concepts were discovered in maize before they were observed in any other species. Maize has biological characteristics that make the plant ideal for many facets of genetic research. An individual plant develops separate structures for reproduction—that is, the male flower (tassel) and the female flower (ear). This reproductive feature allows for easily conducting controlled pollinations, each of which results in 400 to 500 or more kernels each containing an embryo. Thus, one controlled pollination, representing a cross or a self-pollination, can yield a huge number of progeny for genetic analysis. A tremendous amount of data has been accumulated over many years involving the plant's genetics, biochemistry, and physiology.

One of the first and more notable genetic concepts elucidated through the study of maize was evidence that crossing over observed in chromosomes was an actual physical exchange of chromosome parts. In other experiments, it was shown that the events of chromosome crossing over and mutation could occur anywhere within a gene. With regard to higher forms of life, this genetic fine-structure analysis was first demonstrated in maize. Controlling elements (transposons), sometimes described as "jumping genes," were initially detected in maize. This was an astonishing discovery made by Barbara McClintock of genetic fame. These transposons are pieces of DNA that have the ability to move around in the chromosomes of the organism like nomads and affect gene expression. McClintock was aware of this surprising concept of genetics in the 1950s, long before transposons were shown in other species. Maize was Barbara McClintock's life. In 1983, she was awarded the Nobel Prize in Physiology and Medicine, a very deserved award.

Beautifully colored ears of maize, all due to genetic mutations, provide us with ornamentals. Corn stalks are everywhere at Halloween time. Corncobs make pipes for smoking. Maize mazes are even becoming popular for recreation.

Amazing maize is much more than a catchy phrase.

FURTHER READING

Federoff, N., and D. Botstein. 1992. *The Dynamic Genome: Barbara McClintock's Ideas in the Century of Genetics.* Cold Spring Harbor, NY: Cold Spring Harbor Laboratory Press.

Keller, E. F. 1983. *A Feeling for the Organism: The Life and Work of Barbara McClintock.* New York: W. H. Freeman.

Kiesselbach, T. A. 1999 (Reprint). *The Structure and Reproduction of Corn.* Cold Spring Harbor, NY: Cold Spring Harbor Laboratory Press.

Neuffer, M. G., E. H. Coe, and S. R. Wessler. 1997. *Mutants of Maize.* Cold Spring Harbor, NY: Cold Spring Harbor Laboratory Press.

Sheridan, W. F. 1982. *Maize for Biological Research.* Charlottesville, VA: Plant Molecular Biology Association.

REFERENCES

Agronomy and Plant Genetics. 2000. Minnesota Report 247, 63–70.

Creighton, H. B., and B. McClintock. 1931. "A Correlation of Cytological and Genetical Crossing-Over in *Zea mays.*" *Proceedings of the National Academy of Sciences* 17: 492–97.

CHAPTER 17

SEEDLESS WATERMELONS

Most higher organisms have their chromosomes in pairs. The two chromosome members of a pair are described as being homologous to each other; that is, they are not usually identical to each other but they do show a fundamental similarity with regard to structure and the same linear sequence of alleles. The condition of chromosomes in pairs is considered diploid. Some organisms, especially plant species, have more than two homologous chromosomes, a condition called polyploidy (many homologues per chromosome type).

All chromosome conditions beyond pairs are described as polyploidy. Consider A, B, and C to be three different chromosomes in an organism. Then,

Homologous Chromosomes	Genetic description	Chromosome constitution
2	diploid	AA BB CC
3	triploid	AAA BBB CCC
4	tetraploid	AAAA BBBB CCCC
6	hexaploid	AAAAAA BBBBBB CCCCCC
etc.		

Geneticists estimate that 40 to 50 percent of all plant species are now polyploid. Wheat, for example, is a hexaploid containing six sets of chromosomes. The polyploid condition can also be seen in some lower forms of animal life and even in a few species of amphibians and fish. In these cases, polyploidy has become normal for the species. In addition, however, polyploidy can occur erroneously in any species that is normally diploid. Most often such a polyploid occurrence in the progeny of a normally diploid organism will abort. Unlike plants, higher forms of animals cannot tolerate polyploidy very well. Occasionally, an exception to this generalization occurs; for example, a triploid chicken or some other polyploid vertebrate might happen. But it is a very weak, sickly chicken, and such situations are rare.

When polyploidy is normal within a species, the chromosome condition is almost always even numbered—that is, tetraploid (4N), hexaploid (6N), octaploid (8N), and so on. The underlying reason is that sexual reproduction requires homologous chromosomes to equally segregate into sex cells to prevent chromosomal unbalance. A triploid organism, for example, could not readily produce this equal distribution of chromosomes since it has three of each chromosome instead of two, four, six, or eight. Thus the sex cells, or the progeny, from a triploid (3N) organism or a pentaploid (5N) organism will usually abort. In other words, the organism can be regarded as being almost completely sterile and unable to generate seeds.

Every once in a while a geneticist will seize upon basic genetic information and devise an application for general public use. Such is the case with seedless watermelons. This particular researcher had both viable diploid (2N) and tetraploid (4N) watermelon plants available to him. Thus, the cross between these varieties resulted in progeny that were triploid (3N).

Parent plants:	4N (×) 2N
Sex cells:	2N 1N
Progeny	3N (triploid)

Being triploid makes the task of producing genetically balanced sex cells extremely difficult. The reason is due to the many combinations possible when there are three homologues for every chromosome in the organism. For example, consider three homologues for each of only five different chromosomes.

Triploid for five chromosomes:

```
_____     -----     =====   *****   >>>>>
_____     -----     =====   *****   >>>>>
_____     -----     =====   *****   >>>>>
```

The probability of getting genetically balanced sex cells following their segregation in meiosis is very low.

```
_____     -----     =====   *****   >>>>>
or
_____     -----     =====   *****   >>>>>
_____     -----     =====   *****   >>>>>
```

Almost all sex cells from a triploid (three sets of chromosomes) will end up with one of some chromosome types and two of other chromosome types simply owing to probability. Such sex cells are chromosomally unbalanced and not viable, and therefore mature seeds cannot develop. The very small white seeds observed in the seedless watermelon are actually the result of abortions due to the unbalanced chromosome conditions. However, a mature seed will occasionally be found in this more expensive watermelon. It is possible to get a chromosome balance in the sperm cell (but rarely) and a chromosome balance in the egg cell (but rarely), and then have these two cells get together for fertilization (but very rarely). All of these events occurring together would have a probability of rare times rare times rare, simply meaning it will not often happen. Every once in a while a geneticist will seize upon basic genetic information and devise an application for use by picnickers and the general public.

Why are these watermelons more expensive than the other melons? Crosses have to be made between diploid and tetraploid plants to produce seedless watermelons because the triploids do not have seeds. One cannot eat a seedless watermelon on a hot day without thinking about the clever genetics involved.

FURTHER READING

Kihara, H. 1958. "Breeding of Seedless Fruits." *Seiken Ziho* 9: 1–7.

REFERENCES

Burnham, C. 1962. *Discussions in Cytogenetics.* Minneapolis: Burgess, 251–70.
Kihara, H. 1951. "Triploid Watermelons." *American Society of Horticulture and Science Proceedings* 58: 217–30.

CHAPTER 18

EATING GENETICALLY MODIFIED FOODS

G MOs are genetically modified organisms. They are organisms produced by genetic engineering methods rather than traditional breeding methods. Conventional breeding brings chromosomes together from two parents. Genetic engineering usually adds only one or a few genes to the host organism, and such genes are called transgenes. The host organism will hopefully keep all of its normal characteristics except for the single modification brought about by that one gene. Traditional breeding also requires species compatibility with regard to sexual reproduction. Genetic engineering allows for gene transfer from one species to another very different species and at great speed. For example, a gene from a fish (flounder) has been placed into tomatoes to prevent the tomatoes from freezing. You can't cross a flounder with a tomato plant, but GMO biotechnology is powerful, and it can bring about this genetic combination.

Many of the GMOs that have been developed thus far are in the plant world. It has been estimated that 70 percent of processed food in the United States contains products of genetic engineering. Two major types of traits are being genetically engineered into agricultural plants: (1) resistance to insects, and (2) resistance to herbicides. Corn, cotton, and potatoes are examples of genetically engineered plants resistant to insect pests. Corn, soybeans, and canola are examples of genetically engineered plants resistant to herbicides.

Crops resistant to insects had a gene placed in them that is naturally found in the bacterium *Bacillus thuringiensis*. Genetically engineered corn with this bacterial gene is therefore called BT corn. The gene codes for a toxic protein that is insecticidal. The protein coded by this gene kills insects by binding to cells in the midgut of the insects, causing these cells to burst. The plant, therefore, can make its own powerful pesticide against certain insect species. The scheme apparently works well against corn borers and earworms. Cotton and potato plants are also protected against certain insects in this way. In other cases, herbicidal-resistant crops allow farmers to use substantial amounts of herbicides without causing harm to their crops. They don't even have to cultivate the crops. More than half of the soybeans grown in the United States have resistance to the herbicide *Roundup*. The Arctic flounder gene has been placed into strawberries for the purpose of making them more cold resistant. Other attempts are being made to make tomatoes tolerant to salt water so that this crop could actually be grown with seawater. And then there is "golden rice." This rice contains two daffodil genes and one bacterial gene. Together these genes code for metabolism leading to vitamin A production, hence the golden (yellow) appearance of the rice.

The advantages of genetically engineered crops have been consistently outlined for the public. Less use of insecticides means less pollution and less cost to the farmer. Prevention of weeds results in a better harvest. Less freezing translates into more tomatoes and strawberries and less waste. And what is wrong in getting more vitamin A from eating rice, a staple foodstuff eaten by much of the world's population? The breakthrough should drastically reduce vitamin A deficiencies, which lead to night blindness and other nutritional diseases.

Genetic engineering aimed at GMOs is teeming with activity. Not too many decades ago scientists knew very little relative to what genes are all about. Then James Watson and Francis Crick discovered the structure of DNA in 1953. However, the secrets of how the molecule worked still needed to be elucidated. Following those eventual discoveries, the genetic code was spelled out—literally. Researchers next began to unravel some of the genes' control mechanisms and to obtain a better understanding of the nature of mutation and how organisms differentiate. By the 1970s, people were already recalling Aldous Huxley's *Brave New World*, with its mechanical wombs and predestinations. Some found such genetic advances exciting, but others became frightened of them and very cautious of any genetic advances.

There is a need to carefully describe what is really meant by genetic engineering. A good place to begin is to point out what is *not* genetic engineering—that is, in a purist sense. Amniocentesis, in vitro fertilization, embryo transfer, artificial insemination, sex selection, and cryopreservation of sperm, eggs, and embryos do not constitute genetic engineering. Even the cloning of organisms is not considered genetic engineering. Although all of these techniques may be medical advances and require great skill in some cases, none of them change the heredity of an organism in any way. Genetic engineering in a true sense is the manipulation of genetic material. Examples would be the deletion of certain genes, or the replacement of certain genes, or the alteration of certain genes. In other words, genetic engineering is literally conducting genetic surgery to change the heredity of an organism.

Geneticists interested in conducting genetic engineering have mostly concentrated on transferring segments of DNA from one organism to another. Generally the process includes the following steps:

1. The isolation, or synthesis, of the DNA segment or gene to be transferred
2. The cloning or propagation of the DNA segment
3. The transfer of the DNA segment into the host cell or organism
4. The stabilization of the DNA segment in its new genetic environment

Scientists have, for the most part, overcome the problems associated with these tasks when working with one-celled bacteria. However, some technical difficulties still remain when applying the techniques to higher forms of life. DNA segments, often composed of specific genes, can be isolated from organisms and purified. Once a DNA segment has been isolated, it can be replicated to give the researcher many copies for use in additional genetic engineering steps. Often the replication step is accomplished by recombinant DNA techniques that allow DNA from various organisms to be joined. For example, DNA from any organism can be joined with plasmids of bacteria. Plasmids are little closed circles of DNA that exist in addition to the chromosome of the bacterium. After joining foreign DNA to the plasmid, it can then be returned to the bacteria. When the bacteria multiply, the plasmids and their passenger DNA are also multiplied. The multitude of copies is then recovered from the bacteria through straightforward molecular techniques. The procedure is called gene cloning, not to be confused with the cloning of organisms.

Genetic engineering also requires some way to transfer the gene or genes in question into the cells of another organism. This step can be the most difficult to accomplish. A number of different methods have been used:

1. The incubation of host cells with the purified DNA segments to be transferred
2. The injection of the DNA segment directly into the nucleus of the host cells with extremely fine microinjection tools
3. The fusion of cells to transfer whole chromosomes from one cell to another
4. The use of a special gun, called a biolistic gun, that literally shoots the DNA into the host cells, much like a .22-caliber rifle
5. The use of vectors—including plasmids, viruses, and even tiny structures called liposomes—to carry the DNA into host cells

The final requirement for successful genetic engineering is that the newly inserted gene or genes function in a stable and normal manner. Most genetic techniques allow little control over where the gene will be incorporated into the host DNA, and how it will behave after such a transfer. A gene may express differently depending upon where it is located within the host cell's chromosomes. This phenomenon is called position effect. Often, this last step requires the crossing of fingers.

Not everyone, however, is overly joyous about genetically engineered food, sometimes called "Frankenfood." People are asking whether these foods are really going to be beneficial, or whether it is just more big-time agricultural business. Some are concerned about whether these foods are actually safe to eat, and if OK today, will there be any long-term effects? Some cultures don't like the idea of eating odd-looking rice. The GMO rice is yellow instead of white. Another question is whether some people might be seriously allergic to genetically engineered foods; hence, the call has been made for labeling such foods as being GMOs. So far, such allergies have not shown up. The possibility that crops making their own insecticide might have an effect on nontarget insects has probably received the greatest amount of public interest. Some studies have indicated toxicity to moths and butterflies. Other studies of normal field conditions have indicated that this risk has been exaggerated. Again, long-term effects are not yet known. Still another possible problem is whether these transgenes in crops can spread to some of their relatives—for example, creating weeds resistant to herbicides or weeds resistant to insects. Most studies thus far show such gene flow to be minimal. In addition, there is always the possibility of creating resistance in insects and noxious weeds simply by natural selection. Resistance is a phenomenon that always seems to occur when given a chance. Continual heavy use of herbicides on herbicide-resistant crops could ultimately bring about herbicide-resistant weeds through natural selection. And insects could become resistant to BT plants and their home brew of insecticide. Battling resistance is a war that never ends whether due to genetic engineering or not. Despite all of these concerns, farmers are increasing their plantings of biotech crops. They like the use of fewer chemicals and the cost effectiveness of gene-altered crops. It doesn't appear that GMOs are going away.

Figure 18-1. GMO spinach. GMO foods are a favorite subject of cartoonists. www.CartoonStock.com. Reprinted with permission.

FURTHER READING

Buttel, F. H., and R. M. Goodman, eds. 2001. *Of Frankenfoods and Golden Rice.* Madison: Wisconsin Academy of Sciences, Arts and Letters.

Teitel, M., and K. A. Wilson. 1999. *Genetically Engineered Food: Changing the Nature of Nature.* Rochester, VT: Park Street Press.

REFERENCES

Chang, S.-B., and H. de Jong. 2005. "Production of Alien Chromosome Additions and Their Utility in Plant Genetics." *Cytogenetics and Plant Breeding* 109: 335–43.

Halford, N. G. 2004. "Prospects for Genetically Modified Crops." *Annals of Applied Biologists* 145: 17–24.

Mascia, P. N., and R. B. Flavell. 2004. "Safe and Acceptable Strategies for Producing Foreign Molecules in Plants." *Current Opinion in Plant Biology* 7: 189–95.

Sonnewald, U. 2003. "Plant Biotechnology: From Basic Science to Industrial Applications." *Journal of Plant Physiology* 160: 723–25.

CHAPTER 19

GROWING HUMAN HORMONES ON THE SOUTH 40

S ome companies are already at work transferring human genes into crop plants—that is, genes that produce disease-fighting drugs, antibiotics, contraceptives, cancer-fighting substances, and so forth. It may be possible someday to have edible vaccinations. Such biotechnology is often referred to as molecular farming. A scientist many years ago pointed out that the day will come when human growth hormone will be grown on the south 40. Scientists are rapidly approaching those predictions. Recall the phrase "You can't get blood out of a turnip." We are no longer so certain about that.

I have never seen the word "DNAology" in print. Nonetheless, it may be an appropriate term. Geneticists have learned how DNA can be manipulated and studied in numerous ways since the famous Watson and Crick showed how the molecule is put together. Clever manipulations of DNA have brought about many ramifications of DNA tampering. Many of these techniques cause some of the public to be apprehensive and even worried about what scientists are capable of doing with this molecule that underlies all heredity. A deluge of such controversial subjects is upon us.

Activity within the realm of DNAology greatly accelerated when scientists learned how to join DNA molecules from diverse sources in an in vitro procedure. In vitro means that the experiment is accomplished in glass, such as beakers, flasks, test tubes, and Petri dishes, as opposed to the activity actually taking place in an intact animal, plant, or microbe. This latter activity is called in vivo. These techniques, referred to as recombinant DNA technologies, have played a huge role in bringing us to the point whereby we may actually produce numerous important substances in crop plants on the south 40.

One of the key discoveries was finding certain enzymes in bacteria and other microbes that can cleave DNA in a very specific way, and as it turned out in an applicable way. These enzymes are called restriction endonucleases because the point at which they cut DNA is so restrictive. The enzymes recognize short segments of the As, Gs, Ts, and Cs (subunits of DNA) having a specific sequence. Endonuclease refers to making the cut within the interior of the DNA molecule. Today, scientists have an arsenal made up of literally hundreds of different restriction endonucleases for molecular studies, each recognizing a different short DNA sequence.

Endonucleases have played a large role in revolutionizing molecular biology. Their usefulness is due to the specific manner in which most of them cleave DNA. Each type of enzyme recognizes a specific DNA sequence and then makes two cuts in this region, one cut in each of the two strands of DNA. However, in most cases the cuts are not directly across from each other (figure 19-1). But there is more to the uniqueness of the cuts. Most of these DNA-cutting enzymes recognize sequences with rotational symmetry, also known as palindromes. Palindromes spell out the same sequence regardless of being read forward or backward. Youngsters named Otto can probably spell their name both forward and backward sooner than youngsters named Ignatius can; or even Jack (everything else being equal). Otto is a palindrome. Someone once came up with the idea that the first palindrome ever uttered was on the occasion of Adam introducing himself to Eve when he exclaimed: "Madam I'm Adam."

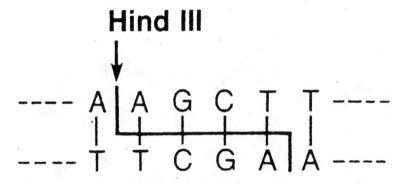

Figure 19-1. Hind III is one of many restriction enzymes available to cut DNA at specific sites. Many of these enzymes, like Hind III, cut the DNA at two different sites on opposite strands.

Palindromes in DNA, however, take into account the two strands of DNA. Reading one strand in one direction is identical to reading the other strand in the opposite direction. Usually these palindrome sequences are four to eight subunits long, with the majority of them being six subunits long. For example, an enzyme called ECO R1 will recognize and specifically cut the DNA wherever it comes into contact with the sequence shown in figure 19-2a. The arrow denotes how the DNA is cut by the endonuclease ECO R1. Note that the sequence of the top strand from left to right is G A A T T C, and the sequence of the bottom strand read from right to left is also G A A T T C. This sequence shows rotational symmetry, and the ECO R1 enzyme will only cut the DNA at such places; that is, the enzyme has this very specific restriction. The cut with ECO R1 would result in DNA having overhangs at the ends made up of A A T T, and T T A A, as shown in figure 19-2b.

The single-stranded sequences making up the overhangs are sometimes called sticky ends; the reason for "sticky" is that T T A A ends will always seek out and bind with A A T T ends, should they ever get close enough to each other. It makes no difference whatsoever whose DNA we are considering. DNA is DNA is DNA. It is possible

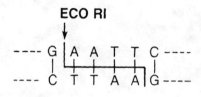

a

b

Figure 19-2. (a) The DNA sequence shown is cleaved by the restriction enzyme ECO R1. (b) Following cleavage, the ends of the DNA molecule have overhangs that are chemically complementary to other DNA molecules with the same overhangs. These overhangs on the molecule can rejoin with other like molecules or with DNA from other sources.

to connect DNA from a duck and DNA from crab grass if complementary ends in both of the DNA segments in question are generated—using restriction endonucleases. This concept is what recombinant DNA is all about. Figure 19-3 shows how a gene from an organism can be cut away from its original chromosome and placed into a bacterial plasmid, a small circle of DNA found in some bacteria.

Since recombinant DNA (recDNA) is generated in vitro (outside the cell in a glass container), geneticists next have to get the DNA back into another cell. Punching recDNA into bacteria became a relatively easy task, because many bacteria have plasmids. These are little closed circles of DNA that exist in addition to the bacterium's chromosome; hence, plasmid DNA is called extrachromosomal. The plasmids can be constructed to incorporate one or

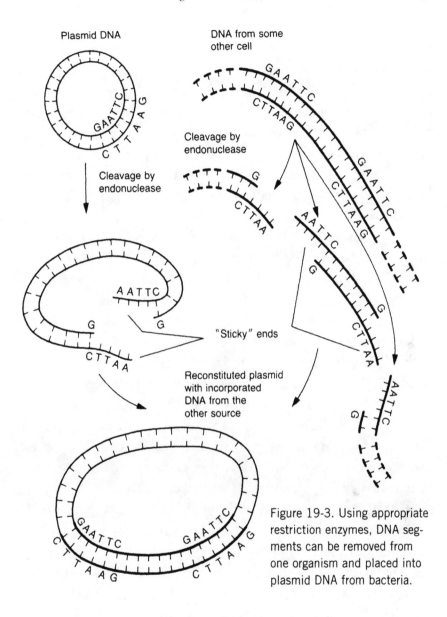

Figure 19-3. Using appropriate restriction enzymes, DNA segments can be removed from one organism and placed into plasmid DNA from bacteria.

more foreign genes from some other organism using recombinant DNA techniques. It is fairly easy to isolate plasmids away from the bacteria, reconstruct them by adding foreign genes using restriction enzymes, and then induce the bacteria to incorporate the plasmids back into their cells.

In some cases, genes of critical importance have been placed into bacteria in the hope that the gene will then be expressed in this new DNA environment. If so, geneticists can capitalize on the system by isolating the gene product from the bacteria. No better example exists than the gene for human insulin that has been placed into bacteria. Bacteria, tiny microbial organisms that never suffer from diabetes, churn out human insulin that can be extracted from them. This resultant insulin is called humulin. Heretofore, insulin had to be obtained from mountainous piles of pig pancreases at slaughterhouses. Insulin is a hormone found in trace amounts like all other hormones. And even after isolating the minute amounts of insulin from these pig tissues, the result is still pig insulin, not human insulin. Producing human insulin by bacteria is a great piece of DNAology—actually the result of genetic engineering.

An important question, however, is how do researchers shove genes into organisms other than bacteria, like higher plants and animals? Various methods exist such as cell fusion, the use of an electrical application that makes huge openings in the membrane for the entrance of genes, the use of a particle gun that literally shoots the gene into other cells, the use of microinjection systems, or the use of some other vector to carry foreign genes into the cells. Virus particles are often used as a vector; that is, certain viruses can serve as vehicles to get the gene or genes into other cells.

One of the best methods to transfer foreign genes into plant cells is through the use of the bacterium *Agrobacterium tumefaciens*, which causes crown disease in plants. Crown disease is a tumorlike growth brought about by the infection of the plant by this bacterium. Once the infection occurs, it changes the DNA of the host plant, causing the tumorlike crown disease. The bacterium has a plasmid that it transfers to the plant cells and actually becomes inserted into the plant's DNA. The plant then expresses genes on the plasmid.

Researchers recognized this bacterium as being a useful tool to direct genetic changes in plants. Using recombinant DNA technologies, they were able to insert separated genes from other organisms into the *Agrobacterium tumefaciens* plasmid. The genes inserted were those responsible for producing desirable products. In addition, the researchers engineered the plasmid in such a way that it lacks the genes for causing the plant tumors. Usually, a small number of plant host cells are transformed in this way with the engineered plasmid, followed by selection of the transformed cells, and then the regeneration of the small group of cells into mature plants. The final outcome, if successful, is a plant that is now churning out some desirable product.

Difficulties do occur in this kind of work, but nonetheless genetic engineering has been accomplished in a number of organisms, both plant and animal. The resultant organisms following successful genetic engineering are described as transgenic organisms. One of the first transgenic organisms to really grab everyone's attention was the placement of firefly genes for luminescence into a tobacco plant. Indeed, the plant does glow in the dark. It is doubtful that the glowing plant has much practical application, other than proving that such genetic engineering antics can actually be accomplished. A genetically engineered Atlantic salmon has also been developed. A Chinook salmon growth gene and an ocean pout antifreeze gene allow for the fish to grow four times faster and larger than normal. A very useful piece of genetic engineering might be the development of pigs with a human gene for making a blood-clotting agent needed for type A hemophiliacs. Mature female pigs could produce this protein in their milk, and not many pigs will be needed to meet the entire world demand for this clotting factor. And then there is ANDI, the genetically engineered Rhesus monkey. Read the word backward. ANDI stands for inserted (I) DNA. The primate now contains a jellyfish gene called the green fluorescent protein (GFP). The expression of the product from this gene is a green fluorescence. However, the little monkey does not show much green fluorescence, just a small amount of glow in the fingernails and shafts of some hair. But ANDI is indeed a transgenic organism. Numerous other genetic engineering examples exist.

Figure 19-4. Growing important human substances in plant organisms.

The author vividly recalls attending a presentation a while back, in which the speaker discussed genetic engineering. During the question-and-answer segment following the talk, someone in the audience vehemently asserted that it was downright reprehensible to place genes from one species into a different species. The interesting part of this voice against genetic engineering is that this person consistently used humulin for his diabetes. Recall that humulin is insulin produced by bacteria harboring an inserted human gene. It all depends, doesn't it?

FURTHER READING

Emery, A. E. H. 1984. *An Introduction to Recombinant DNA.* New York: John Wiley & Sons.

Richards, J., ed. 1978. *Recombinant DNA: Science, Ethics, and Politics.* New York: Academic Press.

Watson, J. D. 1968. *The Double Helix.* New York: Atheneum.

REFERENCE

Johnson, I. S. 1983. "Human Insulin from Recombinant DNA Technology."
219: 632–37.

CHAPTER 20

GENE THERAPY— CHANGING YOUR HEREDITY

Gene therapy in humans is becoming a reality because progress is definitely being made. The promise of gene therapy revolves around the replacement of dysfunctional genes in people with useful genes. The ultimate result of the transfer of genetic material is to cure a disease or to at least improve the clinical status of the patient. The greatest promise comes from situations in which a specific tissue needs to be genetically changed or "fixed." In such cases, one strategy is to remove the tissue from the body, make the appropriate genetic change, and then replace the tissue into the patient. Bone marrow lends itself well to such a "remove-repair-return" technique. Some success has already been shown in mouse models with sickle cell anemia. Scientists have been able to introduce the appropriate functional gene into the bone marrow of the mouse. We may eventually see this scheme applied to other tissues and organs, like the pancreas or liver. Even the germ cell line, sex cells, zygotes, and embryos in early development are possible targets in these gene therapy endeavors.

Human cells carrying abnormal genes and grown under tissue

culture conditions have been genetically cured in the laboratory. For example, cells showing sickle cell anemia, galactosemia, Lesch-Nyhan disease, and some defective immune diseases have all been changed into normally functioning cells. However, these transformations occurred in the laboratory, in flasks, test tubes, and Petri dishes. Real people are much different from a pile of cells in a test tube, and the challenge is formidable.

Removing the tissue, fixing the genetic problem, and then replacing the tissue will not work in many instances. For example, it is obvious that such gene therapy is not feasible for the brain and some other indispensible organs. Consequently, it is necessary to develop a delivery system to these organs; that is, some kind of shuttle is needed. Developing good delivery systems has been one of the difficult steps in gene therapy. The use of a crippled virus has been one of the favorite vehicles for delivering the gene load. The idea in this work is to package the therapeutic DNA within the DNA of the virus, and render the virus such that it can still infect cells but not be capable of pathogenic consequences. Viruses have long been deemed to be good vectors for the delivery of genes, but they are less than perfect. One obstacle is that the amount of their DNA is limited, simply meaning that they are often not big enough to carry the large genes necessary to overcome many of the diseases in question. And then the danger always exists that the crippled virus might not be crippled as much as we would like; hence it could begin to replicate. Still, the adenovirus, which causes the common cold, remains quite popular for some clinical trials. But vectors ideal in all ways have not been found.

Researchers are looking for nonviral ways to transport genes so as to circumvent the obstacles and dangerous aspects associated with viral vectors. Some researchers are investigating a procedure called lipofection. In this case, the DNA is packaged into small cellular structures called liposomes. When applied to the tissue, it is hoped that the target cells will allow these minute particles to enter. In other experiments, researchers simply expose the tissue to "naked"

DNA and then hope that the cells will incorporate the DNA molecules. Recently, transposons have received some attention as a delivery system. Transposons are mobile pieces of DNA found in certain organisms. Transposons can be made to incorporate the corrective gene (DNA) and then they will insert themselves into the patient's genome (DNA). The gene gun that literally shoots genes into cells has mostly been used in plant biotechnology. Another technique being researched is to inject naked DNA into the bloodstream. The hope here is that muscle tissue might take up the foreign genes and activate them to effect their expression in the muscles. Such a technique may also be effective someday for heart and coronary diseases. Clinical trials are being conducted worldwide at medical centers that make use of many of these techniques.

Hundreds of gene therapy trials have been carried out, or at least they are in progress. In 2003, more than six hundred clinical gene therapy trials were under way. Most targeted maladies are single-gene types and immune diseases. Cystic fibrosis is a prime candidate disease for using the adenovirus as a delivery system. More than thirty thousand people in the United States have this disease, with many deaths occurring in patients twenty to thirty years of age. Mucus clogs the lungs, which become an optimum breeding ground for bacteria. The liposome technique is being tried to insert a copy of the normal gene into affected cells. Some success in this regard has already been shown in animals, and some human trials are in progress. The adenovirus is being explored for gene therapy of sickle cell anemia. The naked DNA therapy is being tried on muscle and liver tissues. The idea of using transposons as a delivery system is being tested against hemophilia in mice, and the mice have shown improved blood coagulation. Other diseases being considered for gene therapy someday include asthma, cardiac diseases, pulmonary disease, stroke, and seizures.

One must, however, be careful of unrealistic expectations. Gene therapy is still highly experimental, and plenty of trials and tribulations happen. Genetic surgeons have had few really good successes in people. Safety is a constant concern, and gene therapy always involves risks. A young volunteer patient died in a gene therapy trial owing to a massive immune response to the adenovirus vector itself. The young man had a nonfatal deficiency of a liver enzyme that was being treated. In other incidents, two of ten children in a French gene therapy trial developed cancer. The children were being treated for severe combined immune deficiency (SCID) using a virus as the vector. SCID can be fatal without bone marrow transplants. In these cases, the gene therapy against the disease actually worked well, but two of the patients developed a leukemia-like condition. If not an accidental and unrelated event, something went awry in which a particular type of cell probably began to proliferate uncontrollably. Such unfortunate events accent the realization that gene therapy problems are very complex.

Regardless of the setbacks, the field is slowly maturing. In fact, scientists are already contemplating gene therapy to the individual while still in utero. The rationale is to correct some types of genetic diseases before any clinical manifestations of the disease can occur. The ultimate objective of in utero gene therapy is to establish another way to produce babies with corrected genes and free of genetic disease. Although challenging, these latter gene therapies are actively being explored in animals, and they have shown some progress. No doubt, much promise concerning gene therapy exists within the scientific community.

Many people have fears concerning this kind of genetic engineering. They feel that the genetic engineers are "playing God," and that their efforts will soon make it possible for parents to design their children or governments to design their citizens. They look toward a quality-controlled society where the popularity of certain genes and the unpopularity of others may blot out the population's essential variability. Various religious groups are sensitive to advances in human biology, and genetic engineering in particular troubles many theologians. They, among others, are opposed to intervening in human evolution with tailored genes, and we hear much concern about whether the researchers, known as gene "splicers," even have appropriate ethics. Certainly the interests of the patient have to be a foremost consideration.

On the other hand, other people feel that there is entirely too much gloom surrounding genetic engineering and that much of the fear is unjustified. They believe that we hear too much about legless humans for space travel, humans with gills for underwater activities, and humans with one pair of hands for heavy work and a second pair for more delicate work. But genetic engineering may offer us great hope that outweighs the philosophical objections. Many people feel that we must proceed with gene therapy. Advocates claim that genetic engineering is just another expression of the essential creativity of humans and that considering the pain and suffering that exists in the world it is time for us to explore these ways to enrich people's lives. Certainly scientists must do the work carefully and cautiously and keep society informed at every step of the way. Special problems are bound to arise, for scientists do not yet know everything about how genes interact with each other. Few would argue that gene therapy still remains on trial. But the techniques could revolutionize medicine. Scientists and the public as a whole will have to show sound judgment.

FURTHER READING

Grobstein, C. 1979. *A Double Image of the Double Helix: The Recombinant DNA Debate.* San Francisco: W. H. Freeman.
Huxley, A. 1932. *Brave New World.* New York: Harper & Row.

REFERENCE

Khorana, H. G. 1979. "Total Synthesis of a Gene." *Science* 203: 614–25.

CHAPTER 21

ANIMALS HAVE STEM CELLS TOO

P lants have stem cells, but so do animals. In normal reproduction, the fertilized egg (zygote) is a totipotent cell. This initial cell has the capability of producing a new organism in its entirety—that is, all of the adult body types and also the specialized tissues needed for the development of the embryo, such as the placenta. Thus far, only a zygote has been shown to be completely totipotent. Within this context, stem cells are animal cells that give rise to many of the 210 different mature cell types in an adult human body. For this reason, stem cells are called pluripotent.

Embryonic stem cells have indeed been shown to have pluripotent capabilities. James Thomson at the University of Wisconsin–Madison discovered these fascinating cells in human embryos in 1998. Stem cells were located in the inner cavity of very young embryos at a stage when the embryo is shaped very much like a small hollow ball. Given the right environment, these cells can develop into many other cell types. Since 1998, numerous researchers have joined in the research of stem cells. Many other groups of people, however, have also become interested in stem cell research, including politicians, patient advocates, ethicists, and religious groups.

When the sperm fertilizes an egg to form a zygote, it immediately begins to divide. After several days, a blastocyst forms. This hollow ball of cells contains a small mass of stem cells surrounded by other cells (figure 21-1). At this point of development, the embryo is only about five days old, and the stem cells are capable of generating many of the specialized cells of a human. However, this capability does not last very long because these cells will soon begin to specialize. It is difficult to effect a switch in specialized cells into different specialized cells.

Many researchers feel that stem cells could lead to enormous clinical benefits and possibly much insight into how humans develop. Researchers want to create many lines of cells that would be self-perpetuating. Medical uses of such stem cells include possible treatments against Parkinson disease, Huntington disease, multiple sclerosis, amyotrophic lateral sclerosis (Lou Gehrig disease), heart failure, Alzheimer disease, spinal cord injuries, and blood disorders. Many major human diseases are due to cell depletion. The strategy is to replace damaged tissue and cell loss by transplantation of stem cells into such tissues.

Stem cell research at this time is not without problems. Sometimes the cells are not easily grown under laboratory conditions. Once the cells are successfully grown, researchers need to coax them into forming the desired cell types for medical use, such as heart

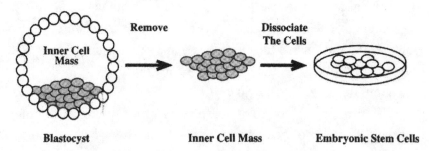

Figure 21-1. The formation and isolation of embryonic stem cells.

cells or neurons. Also, concern exists that the stem cells injected into an individual may form tumors. Still another problem is the possibility of transplantation rejection, even with careful tissue matching. Even if these problems can be overcome, researchers are still concerned about whether stem cells will be able to begin the appropriate gene expression following transplantation.

Mice that are models for Parkinson disease, diabetes, and spinal cord injuries have been successfully treated with stem cells. Expectations that such success will eventually be feasible in humans are abundant. In the past, the number of stem cell lines that researchers could work with was limited. They could only use stem cell lines that were already available and were not allowed to increase the embryo stem cell resources. This moratorium on government-funded research had been reversed, but once again it has been put on hold by the courts. The sources for the human embryos are fertility clinics; that is, they use embryos that would otherwise be discarded anyway.

The controversy arises from the same old question continually asked over and over again for years. When does life begin? This seems to be the wrong question. Any sperm, egg, zygote, embryo cell, or cell of any kind is already *life* if it hasn't already succumbed for some reason. Cellular life doesn't begin. It never ends unless disrupted by outside forces. The question needs to be about when does this cellular life merit being called a human—that is, a human who should be accorded all of the life privileges that you and I now have? This decision can be at any point in the development of the organism that society decides. At any rate, the situation has developed additional strains between science and some religions.

Since embryo cells have proved to be so controversial, many scientists are actively researching adult stem cells. Much hope surrounds adult stem cell research because embryos would not have to be used and destroyed. However, challenges have been put forth regarding their value. Some claim that embryonic stem cells have a greater plasticity than adult stem cells; that adult stem cells are just not as versatile as embryo stem cells. But others think the value between the two sources of stem cells may actually be a tie at this time. For example, bone marrow stem cells appear to be a good source for certain medical therapies like replenishing blood cells destroyed by chemotherapy. Other examples of adult stem cell successes have occurred.

Return your thoughts to embryonic stem cells. Is this small clump of cells in the blastocyst another human being, a new hope for diseased people, or both?

FURTHER READING

Guenin, L. M. 2008. *The Morality of the Embryo.* New York: Cambridge University Press.

Kiessling, A. A., and S. C. Anderson. 2007. *Human Embryonic Stem Cells.* Sudbury, MA: Jones and Bartlett.

Langwith, J., ed. 2007. *Stem Cells.* Opposing Viewpoints Series. Detroit, MI: Thomson Gale.

Ostnor, L., ed. 2008. *Stem Cells, Human Embryos and Ethics: Interdisciplinary Perspectives.* New York: Springer.

Ruse, M., and C. A. Pynes. 2003. *The Stem Cell Controversy: Debating the Issues.* Amherst, NY: Prometheus Books.

REFERENCES

Newsweek. 2001. "The Stem Cell Wars," July 9.

Prentice, D. A. 2003. *Stem Cells and Cloning.* San Francisco: Benjamin Cummings.

CHAPTER 22

STERILE? SO WHAT!

I n the Western world, one in seven people seek treatment for infertility due to a variety of different causes, but often due to sperm quality. Sperm are sensitive cells; consequently, assisted conception is a frequent undertaking. In the United States, 8.5 percent of all couples are infertile due to one or even both of the mating partners. Worldwide the figure for infertility is about 2 to 7 percent. Sperm count in males is also believed to have dropped over the past twenty or thirty years. Therefore, a significant amount of sterility is due to males not producing enough sperm. The idea that "it only takes one" is not quite correct. About fifty million sperm are necessary to make their way to the vicinity of the egg so that one lucky one can enter the egg and pull off a bingo. Motility of sperm has also decreased over this time period. Reasons for these declines are largely unknown.

The reproductive process is so complex that even though it normally proceeds smoothly, it can go awry in many ways. However, genes for good gametes are no longer essential in many cases. Medical science has provided us with a number of ways to circumvent the problems so that people can have babies. They include artificial insemination, in vitro fertilization, embryo transfer, cryopreservation, surrogate mothers, intracytoplasmic sperm induction, gamete intrafallopian transfer, and zygote intrafallopian transfer. However, each of these technological innovations can raise strong feelings and create conflicts among some parts of society.

Artificial insemination (gamete transplantation) is the manual deposition of semen collected from a male into a female's vagina. The semen may come from the woman's husband (AIH) or from an anonymous donor (AID). Pregnancy usual results in 60 to 80 percent of the attempts, and many thousands of children are born each year as results of AIH and AID since the technique is relatively easy to perform. In cases involving a male fertility problem or a desire to avoid passing on a hereditary disorder, the couple may choose AID in order to have a child. An increasing number of single women are also choosing AID. Anonymous donors are often medical students or others deemed to have high intelligence and appearing to be in good health. However, very little information is collected regarding the presence of genetic disorders in donors and their families. When couples are involved, the donors are chosen with an attempt to match the husband's phenotype as closely as possible. Although these selection decisions are genetics at work, we all know that it doesn't always work as expected. Incidences of spontaneous abortion and genetic defects occur no more often than in the general population and maybe even less. Though the technology is similar, AIH is used for different reasons than AID. Some women have used AIH when their husbands were in another country for long periods of time. In other cases, husbands with only a few sperm in their semen have had their sperm concentrated in the laboratory; then AIH has been used to inseminate their wives. Still others have had their sperm frozen and stored before being sterilized, sent to war, or undergoing radiation treatment for cancer.

The freezing process (cryopreservation or cryobanking) is simple and fairly reliable. Collected sperm are treated with glycerol, placed in plastic tubes, and stored in liquid nitrogen at −196 degrees Celsius. Sperm banks can keep sperm stored in this way for many years, and the sperm remain viable to fertilize an egg. The success of this technique has been demonstrated convincingly for years in the livestock industry. However, cryopreservation does require continuous monitoring of the equipment and stored sperm to ensure that the sperm do not warm and deteriorate. Oocytes (eggs) can also be frozen and stored, but the procedure has been more of a technological problem than with sperm. A drastically fewer number of babies have been born using frozen eggs.

There has been some fear that the freezing process required for artificial insemination may damage genes, but no such effect has emerged in the artificial insemination of livestock. The possibility also exists that through widespread use of artificial insemination, an AID offspring may inadvertently marry a close relative. The probability of an AID child marrying a first cousin or closer relative has been calculated as once in about four or five years in the United States. Another potential problem is that donor anonymity may make it difficult for an AID child to find a needed organ transplant. In addition, some couples may be disturbed by a connotation of adultery or illegitimacy, although the courts regard a child conceived by AID with the consent of the husband as his legitimate and legal offspring. AID progeny probably will not know their own chances or their children's chances of developing hereditary disorders. But many of the problems faced by AID children are not unique; similar concerns affect adopted children.

In vitro means "within glassware." In vitro fertilization (IVF) is fertilization outside the body, in a test tube or a Petri dish. Though it has been possible for many years, it has now become more frequently used with many couples. The procedure illustrated in figure 22-1 begins at ovulation when a physician uses a laparoscope, a device equipped with a viewing system and a suction tube that can be inserted through a small abdominal incision to reach the ovary, in order to collect freshly released egg cells. Various drugs can be used to time the ovulation conveniently and to induce multiple ovulations. The collected egg or eggs are then placed in a glass vessel and appropriately treated sperm cells from the woman's husband or a donor are added. If the procedure is successful, fertilization occurs. About twelve hours later, the fertilized eggs (zygotes) are transferred to a second dish. A nutrient medium at this point supports the first

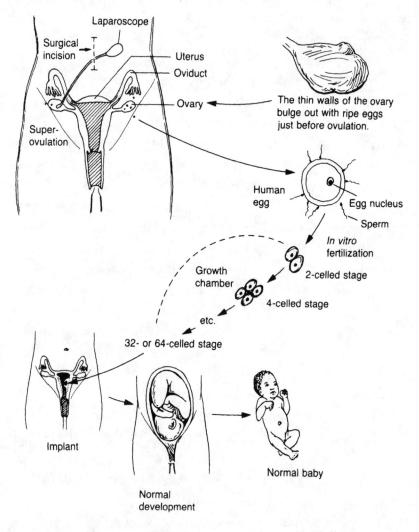

Figure 22-1. The in vitro fertilization procedure. George H. Kieffer, *Bioethics: Textbook Issues*, 1st edition © 1979. Reprinted by permission of Pearson Education, Inc. Upper Saddle River, New Jersey.

few cleavages of the zygote. In about two days, the zygote becomes an eight-celled embryo. Shortly thereafter, the physician collects the embryo with a little of the nutrient fluid and flushes it into the woman's uterus, where, if all goes well, the small embryo implants in the uterine wall. The resultant offspring is often called a "test tube baby" even though the embryo is within glassware for only two or three days of its prebirth development.

P. C. Steptoe, a gynecologist, and R. G. Edwards, a reproductive physiologist, first perfected the in vitro fertilization process. The first birth of a baby by this process occurred in England on July 25, 1978. Her name is Louise Brown. The first American born this way occurred in 1981 and her name is Elizabeth Carr. Both of them were born healthy and without serious genetic problems. Since then, numerous in vitro fertilizations have been performed leading to pregnancies. The success ratio was initially very low, but it has greatly increased over the years.

We can separate sexual activity from procreation with still other manipulations of human reproduction. Both artificial insemination and in vitro fertilization can be used in conjunction with surrogate mothers. In the first case, the surrogate mother is inseminated with sperm from the husband of an infertile couple. The surrogate mother also serves as the egg donor. In the second case, the in vitro fertilized egg is implanted in the uterus not of the egg's source; rather, this female simply serves as an incubator for the embryo/fetus. Transfer of an embryo from the genetic mother's uterus to that of a host mother is another technique that has been accomplished in human beings.

The main reason for the use of surrogate mothers is the prevalence of married women who cannot bear children for one reason or another. Other factors could be the high demand and long waiting lists for adoptees in the United States, or the desire of some women not to interrupt their careers with pregnancies. Cryopreservation of eggs or embryos may also solve this latter concern, allowing for careful planning and timing of births. Cryopreservation of young embryos is possible. A young couple can place a future baby in cold storage and rear him or her later in life. In any event, surrogate motherhood is a novel profession. One motive is surely money, and it has been described as "wombs for rent." Some women have expressed a compassion for those who cannot have children. Still others say that they simply enjoy being pregnant. Wow!

Reproduction can be manipulated in still more ways. Another technique, called intracytoplasmic sperm induction (ICSI), makes absolutely certain of sperm meeting egg. The egg is recovered from the female, and a carefully selected sperm cell is taken from the male and physically injected into the egg using microscopic tools. The resultant zygote is cultured in vitro for a short time and then flushed into the uterus of the female. If the very small embryo embeds in the uterine wall, development can follow. Another unconventional way to make babies is by gamete intrafallopian transfer (GIFT). In this case, several eggs recovered from the female and a significant amount of sperm from the male is placed together in the Fallopian tubes of the female, the usual site of normal fertilization. In still other cases, a zygote from in vitro fertilization can be placed directly into the Fallopian tube. This procedure is called zygote intrafallopian transfer (ZIFT).

Have you ever thought about mechanical wombs (extracorporeal gestation)? After all, science has developed artificial lungs, artificial kidneys, artificial hearts, and rockets to take us to the moon. And what about the use of a chimpanzee or a gorilla as a surrogate mother? It just might be possible, although there could be some difficulty getting the baby away from the surrogate mother.

The bottom line is that if a couple wants a baby, medical technology can accomplish the feat with a fairly good success rate. Many ways are available to bring about those little tax deductions. Something else to think about, however, is whether all of this reproductive manipulation will eventually lead to unwanted changes in the population. In plain words, if various sterilities are due to defective genes, and these people can nonetheless have children through medical advances, those defective genes will be passed on to future generations. This situation then actually opposes natural selection. Think about it.

FURTHER READING

Edwards, R. G., and J. M. Purdy. 1982. *Human Conception In Vitro.* New York: Academic Press.

Huxley, A. 1946. *Brave New World.* New York: Harper and Brothers.

REFERENCES

Biggers, J. D. 1981. "*In Vitro* Fertilization and Embryo Transfer in Human Beings." *New England Journal of Medicine* 304: 336–42.

Cummings, M. R. 2003. *Human Heredity: Principles and Issues.* Pacific Grove, CA: Brooks/Cole–Thomson, 183–85.

Curie-Cohen, M. 1980. "The Frequency of Consanguineous Matings due to Multiple Use Donors in Artificial Insemination." *American Journal of Human Genetics* 32: 589–600.

Kieffer, G. H. 1979. *Bioethics: A Textbook of Issues.* Reading, MA: Addison-Wesley.

Lewis, R. 2008. *Human Genetics: Concepts and Applications.* Boston: McGraw-Hill Higher Education, 413–25.

Steptoe, P. C., and R. G. Edwards. 1978. "Birth after the Reimplantation of a Human Embryo." *Lancet* 2: 366.

Yashon, R. K., and M. R. Cummings. 2009. *Human Genetics and Society.* Belmont, CA: Brooks/Cole, 32–37.

CHAPTER 23

CLONING—
CHIP OFF THE OLD BLOCK

You have undoubtedly heard the saying "chip off the old block." Cloning of humans (or any organism) means exactly that. But the term "cloning" has been broadly used. Technology has developed ways to clone genes—that is, make a lot of copies of a gene from just one or a very few copies of the gene. The resultant genes are identical to each other. Biologists can also clone cells—that is, make a lot of cells from one or very few cells. These resultant cells *should* also be identical to each other.

Cattle have been cloned for some time, but the procedure used is also a much different strategy from how we normally think about cloning. Following artificial insemination, embryos made up of a very few cells are taken away from the cow and literally broken into two or more embryo parts that are then flushed back into other cows simply serving as incubators. Sometimes the embryos are first allowed to undergo further development by implanting them into rabbits. It would probably be wise to remove that developing calf from the rabbit before it is too late. At any rate, the calves born *should* be identical to each other, but not to either the donor of the egg or the donor of the sperm; hence, this cloning is different. The pure sense of cloning refers to the creation of a fully developed offspring by prompting one of the body cells (neither sperm nor egg) to grow, differentiate, and develop into an organism genetically like the cell donor. Having accomplished such a feat, the donor of the cell and the resultant offspring *should* have an identical genetic constitution.

Cloning of humans, or any other animal for that matter, has always been theoretically possible. Now we are actually at this point. The cloning event is possible because of a concept called totipotency. The meaning of totipotency is that single cells have the potential to develop into another organism using the same genetic blueprint that was used previously to develop the organism that donated the cell. To some extent, we see totipotency at work when pieces of roots and other plant cuttings grow into adult plants (called vegetative or asexual reproduction). Some animals can lose a limb and grow it back. Humans cannot do that. But the possibility of growing another "you" from one of your cells now seems to be a realistic genetic event.

Some of the historical events that led us to our present state of the art in cloning need to be briefly discussed. One of the first organisms cloned, in the technical sense as described above, was a wild carrot plant. Researchers grew adult plants from individual cells taken from the stem of the plant by manipulating hormonal applications and the nutrition provided to the cells. Other successes with plants soon followed. Technically, the donor organism and the cloned organism are not parent and offspring; rather, they are twins in a genetic sense. Researchers first cloned frogs in 1952. Because they believed that substances in the fertilized egg's cytoplasm (the entire living part of the cell excluding the nucleus) encouraged totipotency, they replaced a fertilized egg's nucleus with a nucleus removed from a frog embryo cell. Some of these manipulated eggs then developed into tadpoles and even normal adult frogs (figure 23-1). Lack of cloning success or abnormal growth at best, however, resulted if the transplanted nucleus came from a cell of an adult

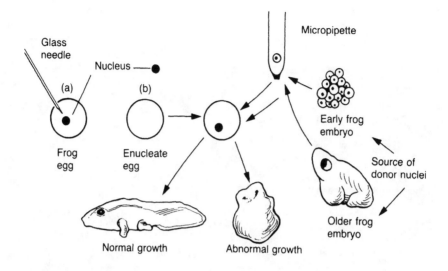

Figure 23-1. The basic procedure of early experiments accomplished with the cloning of frogs. George H. Kieffer, *Bioethics: Textbook Issues*, 1st edition © 1979. Reprinted by permission of Pearson Education, Inc. Upper Saddle River, New Jersey.

frog. By adulthood, apparently too many genes have been permanently programmed in other ways or inactivated. This latter situation has always been the crux of the problem in effecting successful cloning of organisms. Nonetheless, in 1962 a researcher used ultraviolet light to inactivate the nuclei of unfertilized toad eggs. He then transplanted nuclei from tadpole intestinal cells into the enucleated eggs. Some of these manipulated eggs developed into adult toads using the intestinal nucleus to genetically guide their development (figure 23-2).

In these early experiments, two requirements seemed to be essential for any success in the cloning of animals. First, researchers of cloning found that they had to use eggs with cytoplasm that is obviously full of a rich store of messages, signals, and other important growth factors for organism development. Of course, the nucleus of the egg had to be removed first. Second, the nucleus placed into the egg's cytoplasm needed to be from an organism in its

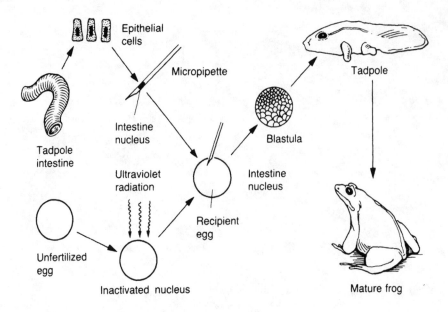

Figure 23-2. The basic procedure of cloning with the use of donor cells not of embryo origin. George H. Kieffer, *Bioethics: Textbook Issues*, 1st edition © 1979. Reprinted by permission of Pearson Education, Inc. Upper Saddle River, New Jersey.

very early development—for example, an embryo cell. Even making the procedure work with the nucleus of an intestinal cell of a tadpole was regarded as somewhat astonishing. Such a cell has already undergone much differentiation. This cloning procedure, therefore, is more accurately called the nuclear transplantation technique. With mammals, however, the normal fertilization process (egg plus sperm) seemed to be obligatory in order to get another organism to develop—until recently.

On March 31, 1978, a book was published that became an instantaneous shocker. The book, *In His Image: The Cloning of a Man*, was written by David M. Rorvik. The theme of the book was a claim that a human had been cloned in December 1976. Almost all of the scientific community did not believe the claim, but shock waves carried all the way to Congress because no laws existed in this regard. Congress called a hearing and brought in experts for consultation. Mr. Rorvik, although invited, did not attend. The story of the book went something like the following. An unidentified millionaire named "Max" built a state-of-the-art laboratory for the purpose of having himself cloned in "land beyond Hawaii," recruited unidentified top-notch medical talent along with an unidentified researcher called "Darwin," used an unidentified surrogate mother called "Sparrow," and had a child created from one of his cells (actually his twin). Yeah, right! The cell supposedly used from Max was kind of secret, but the narrative reveals biopsies taken from liver cells. This choice was probably not a good one since many liver cells are polyploid; that is, they contain more than the normal number of chromosomes. It is very uncertain what that kid would look like with too many chromosomes. Red blood cells were also mentioned in the book for some unknown reason. Another poor choice! Red blood cells in mammals don't even have a nucleus (DNA). Just about everyone now deems this story to be a hoax.

Hello, Dolly! This story is not a hoax. Along came a sheep cloned from a mammary gland cell. The clone was named Dolly after Dolly Parton (Scottish scientists do evidently have a sense of humor). A mammary gland cell is certainly mature and differentiated. Taking into account previous cloning experiments, researchers believed the cloning difficulties encountered were due to a concept called genomic imprinting. This concept means that genes on chromosomes are conditioned and programmed to play a specific role in development, and these genes usually cannot back up and start another development. However, the Scottish scientists found a way to erase the genomic imprinting of the mammary cell, and then using the nuclear transplantation technique they induced the nucleus to begin development from the starting point and into another sheep. Dolly was a complete success, but only after 434 nuclear transfers and 277 developmental trials. Nonetheless, it certainly created excitement, within both the scientific community and the general public. But Dolly grew into adulthood and showed premature arthritis at five and a half years, a relatively early age for arthritis in sheep. Sheep generally do not experience arthritis until about ten or eleven years old. It is only speculation that this arthritis problem was due to cloning since Dolly was already older at the time of her birth. She was "born" from a cell obtained from a six-year-old sheep. This situation may be cause for consideration when undertaking cloning.

Since the Dolly episode, and the publication of the research revealing the details of the technique (the way of science), a number of other mammals have been cloned. In addition to the first success with Dolly, mice, cows, pigs, rabbits, goats, and monkeys have been cloned. Even a domestic cat has been cloned, and now many pet owners want another Fido or Fluffy. It took eighty-seven embryos to get that one kitten. And we already have over seventy million cats in the United States. Some newspaper people creatively referred to the event as "copy cat." Italian scientists cloned a bull and named it Galileo. The cell came from another bull named Zoldo. But the cloning of mammals in Italy was against the law, so the police confiscated the animal, and presumably Galileo was jailed. How fitting is this event? This is not the first time Galileo has been jailed in Italy relative to scientific accomplishment.

The first clone of an endangered species was a gaur, a wild ox native to Southeast Asia. The researchers used a skin cell of the gaur and an enucleated egg from a domestic cow (Bessie), which also served as the surrogate mother. At birth, the baby gaur (named Noah) appeared normal, but it lived only two days for some unknown reason. It was speculated that the death of Noah might have been for a reason unrelated to cloning. However, the procedure many still be a way to save some endangered species. Researchers are now considering other endangered species for possible cloning.

There has been a rush to ban human cloning. The technique represents a much greater technological intervention in reproductive manipulation than, for example, artificial insemination or in vitro fertilization. The technique upsets many people and provokes ethical questions. Many people of all walks of life have warned against the possibility of human cloning. People worry about how to define "parent" for clones, about whether disposing of defective or unwanted cloned embryos is the same as murder, and about the objectives for cloning. Who should or should not be cloned? Everyone could list some people who should be placed high on a list of people who definitely should not be cloned. Might it depersonalize reproduction? Also realize that the procedure could conceivably make males obsolete (figure 23-3). Many people, scientists included, believe there is no good reason for cloning humans; that we should not do it simply because we have the capacity to do it. Thus far, cloning of humans is banned in the United States.

On the other hand, studies of cell cloning and differentiation can provide valuable insights into cellular physiology, growth, and development. This research may improve our ability to cope with cancer, other cellular diseases, healing of wounds, and the regenera-

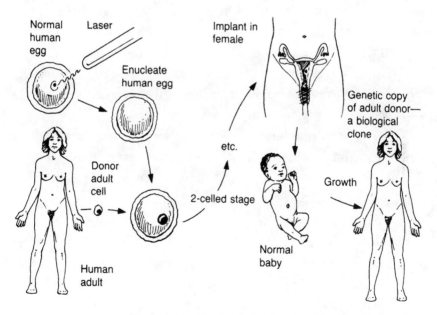

Figure 23-3. The theoretical method for the cloning of humans. George H. Kieffer, *Bioethics: Textbook Issues*, 1st edition © 1979. Reprinted by permission of Pearson Education, Inc. Upper Saddle River, New Jersey.

tion of lost organs and limbs. It may even aid in understanding of aging. And if scientists ever do grow adult clones, they will obviously learn a great deal about the interplay of heredity and environment in the developmental process of the phenotype.

Although there have been some success stories with mammalian cloning, and some reports that the resulting cattle and pigs appear normal, the failures outnumber these successes. Cloning resulting in healthy animals is more difficult than expected. Only 1 to 3 percent of the trials have shown good success. And even then, some of the clones have heart defects, abnormal organs, lung maladies, malfunctioning immune systems, and other developmental problems. Rapid reprogramming may cause errors in the clone's DNA. Still, other results have shown cloned cattle to be perfectly normal, and they did not show any signs of premature aging. But some cloned mice seemed normal at first, and then grew into abnormally obese mice. The problems appear to be genetic issues. In addition, cloned animals, even though they are clones, often have shown large differences from each other with regard to size, health problems, and other appearances. Clones are *supposed* to be genetically identical. Certainly scientists do not yet know everything about reprogramming cells obtained from the donor organism. Still, there are a few people who want to move ahead with human cloning.

The Boys from Brazil is an interesting film. If you have yet not viewed the film, you might want to place it on your "must see" list. The film is based upon a book written by Ira Levin and published in 1976. It is about a Nazi hunter and the mad scientist/doctor Josef Mengele. The story centers on a bizarre plan in which ninety-four clones were developed from Adolf Hitler's cells, taken before he died. The procedure described was amazingly accurate, considering the time of the book and the release of the film. The science of the film, however, did not give much weight to the effects of the environment as these clones developed. The story is fictitious, of course, but very thought provoking.

Figure 23-4. Dolly "born" at six years of age. © 1997 *Star Tribune,* Minneapolis, Minnesota. Reprinted with permission.

FURTHER READING

Kass, L. R., and J. Q. Wilson. 1998. *The Ethics of Human Cloning.* American Enterprise Institute, Washington, DC: AIE Press.

Kolata, G. 1998. *Clone: The Road to Dolly and the Path Ahead.* New York: William Morrow.

McKinnell, R. G. 1985. *Cloning of Frogs, Mice, and Other Animals.* Minneapolis: University of Minnesota Press.

REFERENCES

Briggs, R., and T. J. King. 1952. "Transplantation of Living Nuclei from Blastula Cells into Enucleated Frogs' Eggs." *Proceedings of the National Academy of Sciences* 38: 455–63.

Gurdon, J. B. 1962. "The Developmental Capacity of Nuclei Taken from Intestinal Cells of Feeding Tadpoles." *Journal of Embryology and Experimental Morphology* 10: 622–40.

Kieffer, G. H. 1979. *Bioethics: A Textbook of Issues*. Reading, MA: Addison-Wesley.

Rorvik, D. M. 1978. *In His Image: The Cloning of Man*. Philadelphia: J. B. Lippincott.

CHAPTER 24

AN OLD CONFLICT— NATURE VERSUS NURTURE

F ew would deny that it is not easy to study the inheritance of complex characteristics, such as behavioral traits in humans. Much of what is known comes from extending conclusions reached by means of animal studies to humans. Such extrapolations are accompanied by inevitable uncertainties, but using animals to study the inheritance of behavior does have certain advantages. It is relatively inexpensive to conduct studies using large numbers of rats and mice. They can produce numerous offspring, their environments can be controlled, and inbred (genetically homogenous) lines are available. If animals with different gene forms (alleles) are reared in identical environmental conditions, their behavioral differences can be attributed to genetic causes. If inbred animals are reared in different environments, their behavioral differences can be largely attributed to environmental causes. Naturally, the same observations might be made for humans if we could manipulate them in the same ways.

Geneticists, other biologists, and psychologists, among other groups, have long debated the effects of heredity and environment on human behavior. Except for a few real diehards, most scientists believe that heredity and environment interact to determine many of our human behavioral traits. In fact, most behavioral scientists would probably not argue about whether particular behaviors are hereditary or environmental; rather, they tend to ask whether a genetic component exists in the etiology of the behavior in addition to the obvious environmental components. So where is the nature-nurture controversy? Much argument still remains over the relative importance of the two influences. Those who favor greater hereditary influences and those who favor greater environmental influences can each point to a few clear-cut cases that support their views. Neither group, however, can adequately explain the many behavioral traits that are not clearly controlled solely by heredity or the environment. And therefore, the nature-nurture controversy lives on and on and on like a hundred-year war, or even longer.

The debate is a vigorous one because of its close connection to social issues. Whether behavioral traits such as criminality, intelligence, and alcoholism are due to heredity or environment can subtly, or even blatantly, impact how society approaches them. As an example, the Head Start program was nearly scrapped at one time following a publication in which it was contended that IQ could not really be boosted much through better attention to the environment. Some proponents and opponents have pushed their cases far beyond what the data justify. At one end of the gamut, a few strict hereditarians still exist who think of behavioral traits as being mostly deterministic—that is, largely as a result of genes. At the other end of the gamut, there are some strict environmentalists who think that these traits are infinitely malleable, dependent only upon the environment. Most researchers, however, believe that neither view is accurate. Even if a trait does have a genetic basis, the environment could always affect it, sometimes very noticeably. The environment can modify the effects of genes, and combinations of genes only determine the limits of the range of the phenotypic expression (figure 24-1).

Some behavioral traits in some animals, however, are not deemed genetically complex; rather, the traits are due to one or a few gene pairs that follow straightforward Mendelian genetics. For example, the behavior of bees in their unique societies has always been fascinating to scientists. Bee larvae develop in a comb like a maternity ward under wax. Sometimes, however, the larvae become infected with a *Bacillus* bacterium and they die. Obviously, the bee colony cannot have all of these larvae corpses lying around. So it turns out that in some bee species the population of workers (females) is made up of four different types of bees. Some workers will uncap the wax from over the dead bodies, but then they will not under any circumstances remove the dead from the hive. Some will not uncap the wax, but they will drag the dead out of the hive if others will uncap the wax. Still others have no hang-ups about doing both jobs. Then there are those sensitive ones that want nothing to do with either task. This strange behavior turns out to be the result of Mendelian genetics. Genetic investigations have shown that one pair of genes regulates the uncapping of wax, or not. A separate pair of genes regulates tossing out the dead, or not. If these traits are genetic, specific crosses with specific types of parental bees should result in specific ratios of progeny types that will do particular tasks. And they do! Dragging the dead out of the hive is a behavioral trait, and it has a genetic basis. Animals have been a good source of information in behavioral studies. Animal studies have shown evidence

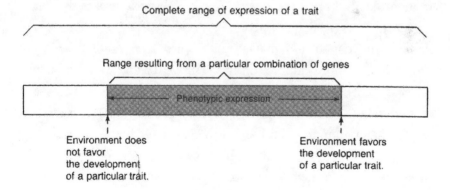

Figure 24-1. Individual genes are discrete, but their contributions determine the limits of a range through which traits are expressed in different environments.

for aggression, sexuality, reproductive strategies, altruism, cooperation, and even rape.

Not every nature versus nurture situation is controversial. Many examples exist in which abnormal behavior is due to specific molecules or to biochemical deficiencies. Tay-Sachs children are deemed mentally challenged within months of being born, and their bodies literally waste away, causing death usually in one to four years. These unfortunate children are missing only one enzyme that you and I have. If not properly treated, children with phenylketonuria will also be severely mentally challenged. They too are missing only one crucial enzyme. And Lesch-Nyhan syndrome is still another disorder whereby a defective gene leads to a mentally handicapped condition, depression, irritability, and even violence. These children are also prone to compulsive self-mutilation. They can chew off their fingers. They can chew off their lips. This behavior can certainly be considered bizarre. If they had hypoxanthine guanine phosphoribosyl transferase (an enzyme) in their cells like you do, they would not mutilate themselves.

Behavioral traits make up a cluster of nature (genetic) versus nurture (environmental) issues. One of the classic ways to study human heredity is the pedigree. This technique is especially useful with traits regulated by a single gene pair. However, most behavioral traits probably have a multigenic mode of inheritance, if they have a genetic basis at all. If a pedigree shows that a trait "runs in the family," the conclusion may point toward a hereditary link, but researchers require more definitive data. After all, family members do share an environment as well as genes. Studies of twins and adoptees have often been used to separate these two factors.

Human twinning always seems to generate a lot of interest, both from scientists and from the public. Twin studies are classical and have been favored as a way to study some traits, especially complex traits. Many researchers still deem the study of twins to be a useful and powerful tool. The twin method capitalizes on the fact that dizygotic twins (fraternal) have half of their genes in common on the

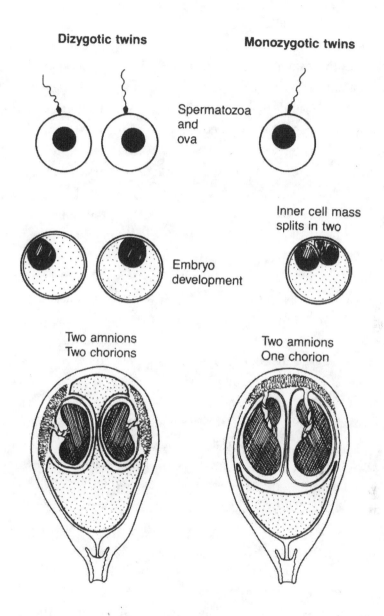

Figure 24-2. Monozygotic twins (identical) are formed when a very early developing cell mass divides into two embryos. Dizygotic twins (fraternal) form when two independently released eggs are each fertilized by a different spermatozoon.

average. Monozygotic twins (identical) actually make up a clone of two, and theoretically have 100 percent of their genes in common. Therefore, the genetic basis of the technique is clear. Monozygotic twins result from a single fertilized egg and, consequently, have identical genes. Dizygotic twins result from two separate eggs fertilized by two separate sperm, and the two twin members will consequently show genetic differences because of meiosis (gamete formation), chromosome combinations, and chromosome crossing over. Recall how the deck of genes is shuffled by these processes. The origin of the two different types of twins is depicted in figure 24-2.

Twins make good research subjects because the genetic situation can help to distinguish heredity from environmental influences, at least to some extent. Identical twins can be reared together or apart—that is, in the same or different environments—as can fraternal twins. Researchers can learn much by comparing the two types of twins in the two situations. With identical twins, any observable differences are assigned mostly to environmental factors. Hence, identical twins reared apart are generally recognized as ideal studies—especially, for example, if one twin member grew up in an apartment on Fifth Avenue in New York City and the other twin grew up on a wheat farm in Montana. One problem is that not many such twin situations exist, at least not as many as researchers would like. Finding a certain trait in both members of monozygotic twins living together does not by itself indicate that the trait was caused by heredity and not by the environment. Such twins living together have identical heredity and nearly the same environment. However, there may be subtle environmental differences beginning as early as their positions in the uterus.

Some time ago, a pair of identical twins had played basketball for a midwestern college. Physically, they were extremely identical due to their identical heredity. They were both very good basketball players, but one clearly had an edge in this regard probably owing to environment. The better basketball-playing member fouled out of a particular game before halftime. In the locker room at halftime, they exchanged jerseys so that the better one could continue playing the game. But some keen pair of eyes knew the difference, and the twins, along with their coach, were dealt severe disciplinary action by the league officials. The physical differences were very subtle, but someone had detected these differences. Maybe it was their mother.

In genetic studies, it is better to use dizygotic twins of the same sex than any other two siblings of the same sex. Like monozygotic twins, dizygotic twins are born at the same time. They are the same age (usually no more than minutes apart), reared at the same time, and generally exposed to the same environment. The situation is not like comparing one sib born when the family was on food stamps to another sib born after oil was discovered in the backyard. It is usually assumed that the environments for dizygotic twin members are just as similar as the environments for two monozygotic twin members. However, this assumption has often been challenged. Some researchers believe that monozygotic twins are often placed in environments that are more alike than those environments experienced by dizygotic twins. Some merit may be given to this criticism. Identical twins are often regarded as being really *special*. Parents tend to dress them the same way, part their hair the same way, and treat them exactly the same way throughout their upbringing. With fraternal twins, one twin member might be treated quite differently than the other twin member. Regardless of these subtle environmental influences, the differences between monozygotic and dizygotic twins relative to certain traits are often striking.

Researchers have a simple way to measure the differences between monozygotic twins versus dizygotic twins. They often express their twin data in terms of concordance rates. Concordance is a term used to describe the situation in which both members of a twin pair exhibit a certain trait. If one twin member has a trait but the second twin member does not have the trait, the pair is considered discordant. The rate is simply calculated as follows:

Concordance = (concordant twins)/ (concordant) + (discordant twins) and presented as a percent

It is not enough to observe only the concordance rate for monozygotic twins. The concordance rate for the dizygotic twins must also be calculated because they do not have identical genotypes. Then a comparison can be made of the concordance rates for the monozygotic and dizygotic twins (table 24-1).

If a trait is highly hereditary, we would expect the concordance rate to be significantly higher for monozygotic twins than for dizygotic twins. If a trait was completely environmental, we would expect the concordance rates to be similar for both kinds of twins.

TABLE 24-1

Two hypothetical examples of concordance calculations

Twin Member Having the Trait	Other Twin Member[a]	Twin Member Having the Trait	Other Twin Member[a]
1	C	1	D
2	C	2	D
3	C	3	C
4	C	4	D
5	D	5	D
6	C	6	D
7	D	7	D
8	C	8	D
9	C	9	D
10	C	10	D

Concordance = 8/(8 + 2) = 80 percent
Concordance = 1/(1 + 9) = 10 percent
Note: [a]C is concordance and D is discordance.

The magnitude of concordance rates is somewhat secondary in importance. For example, concordance rates for measles (before vaccines) in both monozygotic and dizygotic twins were about 96 to 98 percent. These data do not indicate that a hereditary propensity exists for measles. The result is simply a matter of living together, and that the contagious disease is environmental. Death from acute infection has a concordance in both types of twins of about 8 or 9 percent. When acute infection strikes, it seems to be a very environmental situation. Data assembled by researchers, on the other hand, shows manic-depressive psychosis to have a concordance of 67 percent in monozygotic twins and only 5 percent in dizygotic twins. These figures relate to a 13.4-fold difference, and certainly may indicate a genetic component in manic depression.

Human behavior is immensely variable because of the tremendous diversity of human genotypes and the environments in which they reside, both of which play strong roles in molding behavioral phenotypes. These phenotypes include more than one hundred kinds of departure from mental behavior that we usually regard as normal. One of these, schizophrenia, is a mental illness estimated to have a worldwide frequency of about 1 percent (one in every one hundred persons).

Schizophrenia is characterized by delusions, hallucinations, passivity, withdrawal, suspiciousness, and chaotic thoughts. The disorder can be mild, moderate, or severe. However, it is often difficult to draw such lines, or even to distinguish schizophrenia from other types of mental disorders. The hypothetical causes suggested for schizophrenia include both structural and chemical brain abnormalities. One hypothesis indicates an overabundance of the neurotransmitter dopamine, which passes signals from nerve cell to nerve cell in certain parts of the brain. Another hypothesis indicates an overabundance of the cell membrane receptors through which nerve cells respond to dopamine. Still another proposes that a virus or viruslike agent is the cause of schizophrenia. Many investigators are convinced that schizophrenia has a genetic component, but the evidence is mostly statistical.

Schizophrenia certainly runs in families, and a person's risk of schizophrenia goes up with his or her degree of genetic relatedness to a diagnosed schizophrenic. This association could possibly be the result of the relative's mutual environment, which includes one or more schizophrenic persons, and can by itself be a significant factor. However, the concordance data for monozygotic and dizygotic twins in many countries strongly suggest that a genetic component is involved. Although diagnostic criteria for schizophrenia have changed over the years, the general concordance patterns have not changed. Monozygotic twins are more than three times as likely to show concordance for schizophrenia as dizygotic twins. Still, the concordance rate for monozygotic twins only averages about 45 to 50 percent. If the trait was completely hereditary, the concordance rate should be 100 percent. Adoption studies have shown similar patterns. Schizophrenia is significantly more prevalent among persons born to schizophrenic mothers but raised from very early childhood by nonschizophrenic adoptive parents than among control groups of individuals born to nonschizophrenic mothers and adopted by nonschizophrenic parents in very early childhood. Nonetheless, the role of the environment is also evident in these studies. Unfortunately, the specific environmental factors involved in schizophrenia have not yet been definitively identified.

Many behavioral scientists agree that a genetic component exists for schizophrenia, but they can only speculate on the specific mode of inheritance. Some favor a dominant gene model with incomplete penetrance. This means that they attribute the disorder to a dominant gene that is not always expressed. Others prefer the polygenic model that requires schizophrenia to be regulated by many gene pairs. The latter model is probably the more favored of the two. A third model is the genetic heterogeneity model inferring that different schizophrenic persons may have inherited their disorders in different ways. Finally, some researchers think that there may be two completely different types of schizophrenia, one with and one without a strong genetic basis. As one can see, complex traits are difficult to figure out.

Manic-depressive illness occurs in 1 to 2 percent of the population in the United States. Its cause may be similar to that of schizophrenia. Some evidence even suggests that the disorder may be an alternative expression of schizophrenia genes. Environmental factors, of course, may again play a role. Still a strong genetic component can also be demonstrated. As previously mentioned, the overall concordance rate for monozygotic twins is about 67 percent, but for dizygotic twins it is only about 5 percent. In addition, comparisons of adoptees and their biological parents with control groups show a threefold greater incidence of manic depression among people with manic-depressive biological parents. In some cases, stress and other environmental factors may also be necessary for the disorder to be expressed.

Some scientists assert that the link between genetics and alcoholism is well documented. Others are not so sure. Alcoholism is a complex dependence on alcohol with so much variability that it is sometimes difficult to reliably distinguish between alcoholics and nonalcoholics. The disorder (I'll call it a disorder) affects between 3 and 5 percent of all adult males and almost 1 percent of adult females in the United States. The behavioral trait affects every socioeconomic class and people of all levels of IQ and education. Alcoholism is not found simply in "down and out" people. It occurs among successful people in all walks of life. It occurs among businesspeople, politicians, professionals of all types, skilled workers, unskilled workers, and so on.

Alcoholism clearly runs in families, but again this may be due to genetic or environmental causes, or both. Heredity has been implicated by studies of twins, half siblings, and adoptees. In addition, alcohol preference can be bred into rats and mice. Whenever geneticists can breed for something, they begin to think that the trait may have a hereditary component. Of course, humans are not mice or rats. Also, biochemical investigations are contributing meaningful information about alcohol metabolism. In one twin study, severe alcohol abuse among monozygotic twins yielded a concordance rate of 84.5 percent, while the rate for dizygotic twins was 66.7 percent. One adoption study compared men who were separated early from their biological parents, one of whom was an alcoholic, to a control group of adoptees whose biological parents were not alcoholics. The results showed almost four times as many alcoholics among the adoptees with an alcoholic parent. In another study, sons of alcoholic parents adopted in early infancy by nonalcoholics showed alcoholism rates similar to those of their brothers raised by the alcoholic parents.

Genetic vulnerability to alcoholism seems to be a real situation, but everyone having this vulnerability will not necessarily become an alcoholic. The condition is not one of genetic determinism. Instead, scientists suspect polygenic inheritance, with many environmental factors shifting the phenotype one way or the other. Scientists now need to identify the biochemical differences that are inherited. The link between the chemical substance, alcohol, and the actual addiction remains elusive. The data are limited, but since alcoholism is responsible for so much human and social tragedy, the trait should certainly receive continued scientific study.

Do you wonder whether you have "smart" genes? Or did you simply have a super environment? Or both? In other words, is human intelligence controlled at least to some extent by heredity? Some aspects of this question are extremely sensitive, but important because the answer can affect critical policy decisions at all levels of government and education. The problems involved in answering this question, and there are many, begin with the definition of intelligence. Broadly speaking, intelligence is the ability to learn, reason, and create new ideas. But intelligence may be more complex than this broad definition suggests, and no definition seems to satisfy everyone.

Another problem, assuming that we can solve the first one, lies in how to measure intelligence. Most testing schemes ask people to answer questions and solve problems that will produce a score in terms of IQ (intelligence quotient) points. Supposedly, the average score for American whites at any given age by definition is 100. The group of scores from any age group forms a normal distribution curve. On this curve, two-thirds of the population would have IQs within 15 points of 100—that is, between 85 and 115. This distribution appears in figure 24-3. But not all experts believe that our IQ scores really mean intelligence. For example, many do not believe that our IQ tests measure much creativity. Nevertheless, IQ scores do correlate well with other yardsticks of intelligence, such as school performance.

To explain possible associations between IQ and heredity, one needs to understand a statistic called correlation. This statistic is generated by an equation to determine whether an association exists between two traits, or between the same trait in two different groups of people. It is simply a way of checking out in a statistical way whether an association might exist. After pairing the numbers and punching the data into the equation, one ends up with a solitary number. By the very nature of the equation, the number will always be between −1.0 and +1.0. Any result edging toward +1.0 indicates a strong positive correlation. This means that when one variable is a high number, so is the other number paired with it. In other words, an association does exist. If low numbers are paired with high numbers, a negative correlation exists in which the correlation would

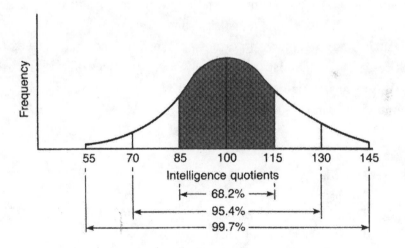

Figure 24-3. The distribution of intelligence quotients among the Caucasian population in the United States.

approach −1.0. If no association (correlation) exists, the value should hover around 0. Correlations, however, do not give any information relative to cause and effect. They just show whether two variables tend to be associated with each other (correlated).

Statistics is one type of ammunition in the arsenal of genetics and many other sciences. Many researchers have calculated correlation coefficients for the IQs of monozygotic and dizygotic twins living together and apart, and for various other related and unrelated pairs of people. With regard to IQ, the correlations are strongest for monozygotic twins living together (0.87) and apart (0.75), and weakest for unrelated persons living apart (0.01). The correlation coefficient for dizygotic twins living together (same sex) was 0.56, and for non-twin siblings living together it was 0.55. The correlation for unrelated persons living together was 0.24, much greater than virtually no correlation of 0.01 for unrelated persons living apart. If correlation coefficients are reliable, these data reveal strong roles for both heredity and environment in how well people perform on IQ tests.

Some people are convinced that we should not even pursue research on these sensitive questions. They contend that the information is too easily misinterpreted and misused, resulting in social injustices. Others disagree just as strongly on the grounds that intelligence is an extremely important human trait, and that it needs to be researched. The nature versus nurture war rages on with regard to this issue and many others. Another such issue, criminality, is presented in the next chapter.

FURTHER READING

Hartl, D. 1977. *Our Uncertain Heritage: Genetics and Human Diversity.* Philadelphia: J. B. Lippincott.

Jensen, A. R. 1969. "How Much Can We Boost I.Q. and Scholastic Achievement?" *Harvard Educational Review* 39: 1–123.

Karlsson, J. L. 1978. *Inheritance and Creative Intelligence.* Chicago: Nelson-Hall.

Parens, E., A. R. Chapman, and N. Press, eds. 2006. *Wrestling with Behavioral Genetics, Science, Ethics, and Public Conversation.* Baltimore, MD: Johns Hopkins University Press.

Penrose, L. S. 1963. *The Biology of Mental Defect.* New York: Grune & Stratton.

Reed, E. W., and S. C. Reed. 1965. *Mental Retardation: A Family Study.* Philadelphia: W. B. Saunders.

Steen, R. G. 1996. *DNA and Destiny: Nature and Nurture in Human Behavior.* New York: Plenum.

REFERENCES

Goodwin, D. W. 1979. "Alcoholism and Heredity." *Archives of General Psychiatry* 36: 57–61.

Goodwin, D. W., F. Schulsinger, L. Hermansen, S. B. Guze, and G. Winokur. 1973. "Alcohol Problems in Adoptees Raised Apart from Alcoholic Biological Parents." *Archives of General Psychiatry* 28: 238–43.

———. 1974. "Drinking Problems in Adopted and Non-adopted Sons of Alcoholics." *Archives of General Psychiatry* 31: 164–69.

Gottesman, I. I., and J. Shields. 1966. "Schizophrenia in Twins: 16 Years' Consecutive Admissions to a Psychiatric Clinic." *British Journal of Psychiatry* 112: 809–18.

———. 1973. "Genetic Theorizing and Schizophrenia." *British Journal of Psychiatry* 122: 15–30.

Hook, E. B. 1973. "Behavioral Implications of the Human XYY Genotype." *Science* 179: 139–50.

Risch, N., and D. Botstein. 1996. "A Manic Depressive History." *Nature Genetics* 12: 351–53.

Rothenbuhler, W. C. 1964. "Behavior Genetics of Nest Cleaning in Honey Bees IV: Responses of F_1 and Backcross Generations to Disease Killing Brood." *American Zoologist* 4: 111–23.

CHAPTER 25

THE BAD SEED

A possible genetic predisposition to criminality is still another nature versus nurture issue. Some argue that genetic and familial elements, rather than just social conditions, play a significant role in predisposition to antisocial behavior. This contention especially involves serious crimes and repetitive criminality. Such a stance, of course, is highly controversial.

Let's begin with a long-ago research report about the infamous Jukes family published around 1916. This family had a terrible track record. Considering 2,096 family members traced over six generations, the record showed 140 hardened criminals, 300 sexually immoral persons, 310 paupers, 7 murderers, 60 common thieves, and 130 members convicted of other assorted crimes. Every grade of viciousness was observed throughout this family. In contrast, 1,300 members of the Edwards family, beginning with the president of Princeton University, were traced in the same way. The record showed 100 clergymen, 95 lawyers, 295 college graduates (at a time when college graduates were quite rare), 65 college professors, 60 doctors, 60 authors, 80 public officials, and 30 judges. No member was ever convicted of a crime. The obvious question would be whether these family differences are indicative of heredity or environment. A case can be made for either factor. One would certainly have to conclude that these behaviors run in the family. However, no definitive data exist to conclude that either of these contrasting situations is actually due to heredity or the environment. The take-home message is that one must be careful in making firm conclusions based simply on familial patterns in analyzing such situations, especially complex traits. Better evidence is needed. Individuals making up families share comparable genes *and* environmental factors.

The quest for better data ultimately leads behavioral scientists back to twin studies. In some studies, monozygotic twins showed a 72 percent concordance rate for criminal behavior, while dizygotic twins showed a 34 percent concordance rate. Such data lead some researchers to support the possibility of a genetic predisposition for criminality, at least to some extent. The idea of a "bad seed" has been around for a long time. But "bad seed" does not infer a single gene pair responsible for such a behavioral trait—that is, a trait acting like it was straightforward Mendelian in its action. If some kind of an association exists between criminality and heredity, the mode would probably be polygenic inheritance (many gene pairs) of various behavioral traits, all of which could further lead to an inclination toward antisocial behavior. But such hypotheses regularly meet with strong opposition.

Some readers, in the very mature age group, may remember the entertainment production titled *The Bad Seed*, first as a Broadway play and then as a 1956 movie. "Rhoda" was a grade school child and thought to be a well-mannered perfect little lady; her only personality flaw turned out to be her willingness to kill people. First she did in an elderly lady she knew, then a grade school classmate, and finally the apartment custodian who teased her. The plot, as you might imagine, revealed a murderer in the family tree, and the "bad seed" leaped a generation and ended up in sweet little Rhoda. It was only a movie and of course fictional, but it served as another catalyst for thought and controversial views.

Striking numbers have been reported in Danish adoption studies. Children of noncriminal fathers adopted by criminal stepfathers (don't know why) were compared to children of noncriminal fathers adopted by noncriminal stepfathers. The crime rates were average in both cases, suggesting that it makes little difference whether the children were reared in a criminal or noncriminal environment. However, children of criminal fathers adopted by noncriminal stepfathers had a higher crime rate than either of the other two groups. And children of criminal fathers adopted by criminal stepfathers (don't know why) showed the highest crime rate of all of the groups. Heredity may play a partial role, but most researchers believe there may be a strong interaction between heredity and environmental factors.

In the past, many studies on the possible inheritance of criminality had focused on persons with the aneuploid conditions of sex chromosomes—that is, the wrong number of sex chromosomes, such as XYY, XXY, and XXYY. These conditions are called sex chromosome karyotypes. Data had indicated that XYY individuals were thirty to forty times more likely to be placed in penal institutions than XY persons (normal chromosome situation). Those initial studies were criticized for lacking good controls, among other reasons, but the fact remained that a greater proportion of the XYY population than the XY population was found in penal institutions. Other studies revealed that about 2 percent of all male inmates had the XYY karyotype, while the XYY frequency among all male newborns was only 0.1 percent. One can easily see the twentyfold disproportionate situation when making this comparison. At first, some researchers thought that the extra Y chromosome in these individuals may be the culprit causing the antisocial behavior. The Y chromosome always gets a bad rap. Additional data, however, indicated that individuals with the XXY karyotype have similar behavioral tendencies. Such males have an extra X, not an extra Y chromosome. The proportion of XXYY individuals eventually committed to prisons is even more dramatic, although this is a much rarer chromosome condition providing us with very little data.

There is little doubt concerning the frequencies of XYY and XXY persons found in mental and penal institutions. These studies have shown that persons with such chromosome conditions are much more likely to be institutionalized than persons with the normal XY karyotype. The result was just a matter of counting and using arithmetic. Nonetheless, the cause of this relationship is far from clear. No one has ever definitively discovered a gene for crime. The chromosome data, however, have been evaluated in relation to other factors. Persons with these chromosome conditions often show IQs significantly lower than normal. Also, the average IQ of institutionalized inmates is significantly lower than the average IQ of the population as a whole. However, an exception might be the IQs of the Nazi criminals who underwent war-crime trials at Nuremberg. They had IQs that ranged from 107 to 143. Such IQs are fairly decent, but their mental abilities were used in villainous ways. Notwithstanding such minor exceptions, *some* people with below-average IQs have a greater tendency to exhibit above-average levels of antisocial behavior. These observations lead some scientists to conclude that the lower intelligence of antisocial individuals may be the main link in the sex chromosome karyotype–prison association. In addition, XY persons may have better legal council than XYY persons. At any rate, it should be noted that inmates with XYY or XXY karyotypes are mostly guilty of crimes against property, and not so much crimes of violence.

Little doubt exists that the chromosome-based antisocial situation presents some difficulty for society and should not be regarded as trivial. On the other hand, everyone must be careful not to foster unwarranted prejudices by stigmatizing people. It is imperative to realize that most XYY and XXY persons live perfectly normal law-abiding lives. Little would be gained by labeling them at birth or later before they have committed any kind of crime, which was actually a proposal at one time. And by all means remember that if 2 percent of the inmates have these chromosome irregularities, 98 percent of the inmates are just badly behaved XY individuals.

FURTHER READING

March, W. 2005 edition. *Bad Seed.* New York: Harper Perennial.

Mednick, S. A., and K. O. Christensen, eds. 1977. *Biosocial Bases of Criminal Behavior.* New York: Gardner.

Oakley, B. 2008. *Evil Genes.* Amherst, NY: Prometheus Books.

REFERENCES

Dugdale, R. L. 1900. *The Jukes: A Study in Crime, Pauperism, Disease, and Heredity.* New York: G. P. Putnam's Sons/Knickerbocker Press.

Mednick, S. A., W. F. Gabrielli Jr., and B. Hutchings. 1984. "Genetic Influences in Criminal Convictions: Evidence from an Adoption Cohort." *Science* 224: 891–93.

CHAPTER 26

CATCHING BAD PEOPLE AND FREEING INNOCENT PEOPLE

Disputes over biological relatedness often end up in court. Also, the identification of perpetrators of crimes often needs evidence based on genetics. In the past, blood samples were mostly used, but the conclusions were based simply on the antigens on the red blood cells (ABO blood system) or the antigen system on the white blood cells (human leukocyte associated). However, the results only permitted the exclusion of wrongly accused persons. Determining the guilty with a high probability of certainty was not possible. Today we have the more powerful tool called DNA fingerprinting.

DNA fingerprinting has many applications. The Unknown Soldier is no longer unknown. Switched babies in hospitals have been correctly sorted out and given to the right parents. Missing children have been identified and linked to their rightful relatives. Paternity cases have been solved. Thomas Jefferson's sexual activities have been brought to light. Immigration disputes can be scientifically handled. Evolutionary studies are being conducted with some scientific surety. The usefulness of DNA testing, however, goes even further than those applications. Convicting criminals and clearing the innocent are a couple of the most important applications of this genetic breakthrough. DNA fingerprinting (also called DNA profiling) has revolutionized the world of forensics. The application of genetic knowledge and methodology to solving crimes is called forensic genetics. The critical elements of these tests center upon the genetic diversity of individuals.

DNA fingerprinting is kind of a misnomer. The technique, developed by Alec Jeffreys in England, is a creative piece of work. It would make Dick Tracy with his futuristic crime-fighting tools beam with delight. Particular DNA segments that are scattered about in the chromosomes of people are known to vary greatly from one person to another. These regions of DNA are not important for survival, so they can undergo changes over long periods of time without any genetic consequences. The sequence of those A, T, G, and C subunits of DNA are again involved in this procedure. The odds are that no two people have exactly the same repertoire of DNA within these segments, except for identical twins. However, a remote chance always exists that someone somewhere in the world might have the same DNA profile as a suspect, but the probability of such an event is usually one in many millions or even billions. Still it is essential that other evidence would also point to the prime suspect and not to someone else who, for example, lives on the northern edge of Siberia. The materials scrutinized by forensic geneticists are important clues. Violent crimes can leave bloodstains. Rape victims can have bits of their attackers' skin under their fingernails, and in some cases, semen samples may be available.

The following description is the bare basics of this exciting methodology (figure 26-1). First, forensic scientists have to obtain some cellular tissue for extraction of the DNA, such as blood, hair, skin, semen, or saliva. They can then extract and purify the DNA from the available cells in these samples. They do not need much DNA from the sample because it can always be amplified by a technique called polymerase chain reaction (PCR). This procedure actually works on the principle of a chain reaction. One copy becomes two, two become four, four become eight, and so on. Scientists can place a minuscule amount of DNA in an instrument, program it by computer, come back tomorrow, and they literally have millions of copies of the specific DNA to analyze. If the forensic workers can obtain a minimum of one hundred to two hundred cells, they will be able to conduct the procedure. A single hair will sometimes be enough. Next, the forensic scientists cleave the DNA sample using

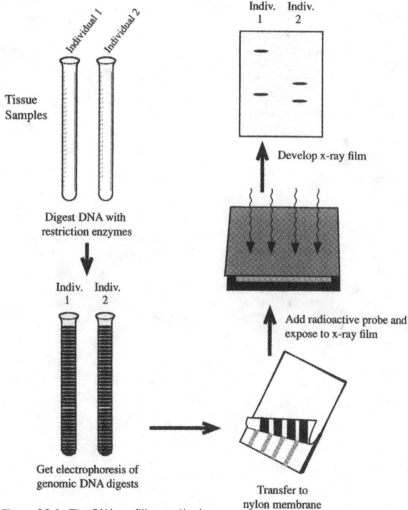

Figure 26-1. The DNA profiling method.

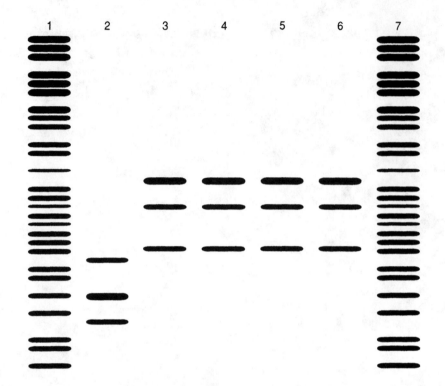

Figure 26-2. Lanes 1 and 7 contain DNA used simply as markers of size. Lane 2 is a control of a known DNA profile. Lanes 3, 4, and 5 are three samples of DNA obtained from the crime scene. Lane 6 is the DNA profile of a suspect. A match is shown.

restriction enzymes that cut the DNA at very specific sites. This step is an important underlying concept of DNA fingerprinting. Everyone's DNA will be cut by these restriction enzymes into pieces of varying sizes. The different sizes, in turn, can be separated by running them through a gel-like material with an applied electric current (called electrophoresis). Once separated, the DNA pieces are transferred to a nylon filter and probed with specific radioactive segments of DNA (called probes) that will bind to the individual's DNA on the filter. Eventually, the entire DNA pattern, appearing like a supermarket bar code, is displayed on x-ray film, developed to make it permanent, and then analyzed (figure 26-2). A popular motto among forensic workers is as follows: "If you *leave* it, we will *cleave* it." Blood, hair, semen, or any other tissues left behind at crime scenes are literal biological calling cards.

A more detailed basis for cutting up everyone's DNA into different sizes is due to repeated segments of DNA. Certain blocks of DNA among our chromosomes are made up of identical DNA segments repeated over and over again, sequentially in tandem. The number of these repeats within a region varies greatly among people (called polymorphisms). Hence, when these pieces of DNA are cut out at their beginning and end, different sizes will exist among people. The technique mostly favored at the present time makes use of short tandem repeats of DNA (called STRs). A number of other techniques have been developed for DNA profiling. Nonetheless, most of the techniques thus far are based upon the phenomenon that certain regions of DNA among people, when cleaved, will result in different sizes. And the resultant pattern identifies you.

Through the media, people are often made aware of the statistics involved in criminal cases using DNA profiling, which is usually staggering. For example, assume that twenty different sizes exist at one specific DNA site. Since we all have two sets of DNA, one set obtained from our mother and one set from our father, the number of combinations for this one DNA site would be 210 (20 + 19 + 18 + etc.). Now assume that four different DNA sites are analyzed, each with twenty different sizes. The number of different profiles possible then becomes 210^4, or approximately two billion. The forensic scientists actually use thirteen different key sites, but the frequency of a certain number of repeats in the population also has to be taken into account. Nonetheless, the probability of a match becomes nearly certain. If enough sites are analyzed, any particular DNA profile becomes very unique.

Forensic DNA profiling has been an immense aid in convicting numerous people for their dastardly deeds of rape and murder. In an early case in Germany, DNA from 16,400 men in an entire region was analyzed. The strategy led to the arrest and conviction of a man for the rape and murder of an eleven-year-old girl. Even old crimes reaching the "cold case" status have been solved if the blood or semen was stored. The high-tech detectives using these techniques are investigating many old, unsolved murders and rapes. One murder conviction was aided by the DNA obtained from fruitlike pods dropped by trees and found in the back of the pickup owned by the suspect. The tree pods were shown to have come from the trees at the murder scene. The DNA profile from these trees was clearly different from one hundred other trees of that same species not growing in the vicinity of the crime.

The other important side of DNA profiling is the large number of persons shown to be innocent, some many years after they were incarcerated. At least one of them had already died in prison. This man was on death row for fourteen years, scheduled for execution for a rape and murder conviction. After he died of cancer, it was shown that he did not commit the crime. Shoddy legal work and unknowing juries have wrongfully imprisoned more than a few persons who are now being released because of DNA testing.

Most people with any knowledge about the accuracy of DNA fingerprinting agree that the technique is the most stupendous advance in forensics since the other kind of fingerprints. However, strict standards must be in place and followed. A DNA profile match does not automatically prove guilt. Although remote, there is always the chance that two people might have the same profile. Hence, prosecutors strive to have other evidence to support the DNA analysis. In addition, laboratories need to be extremely diligent to prevent human error or contamination. Also, a suspect not involved in the crime could conceivably leave DNA at the scene previous to the crime or even after the crime.

Now that the case has been made for DNA fingerprinting as an important technology for solving crimes, would you be willing to give up some of your DNA for a national database? In the United States, we already have a DNA database (DNA Index System) made up of DNA from hundreds of thousands of offenders. But how about having a mass DNA-screening program to place everyone's DNA into the database? Tremendous forensic technique or not, much opposition exists to any plan for mass screening. People fear that the DNA profiles will be used wrongly for insurance purposes or covertly used by employers. Certainly laws are necessary to protect against such DNA misuses. But the database only needs the DNA profile of those variable regions that make forensic DNA finger-printing possible. The entire DNA sample can be subsequently discarded. Some people predict that the day will come when everyone's DNA will be in a national database. *Gattaca* (a must-see movie) might be just around the corner.

FURTHER READING

Lee, H. C., and F. Tirnady. 2003. *Blood Evidence: How DNA Is Revolutionizing the Way We Solve Crimes.* Cambridge, MA: Perseus.

Marzilli, A. 2005. *DNA Evidence.* New York: Chelsea House.

Reilly, P. *Genetics, Law and Social Policy.* Cambridge, MA: Harvard University Press.

CHAPTER 27

FREE WILL OR NOT

Any discussion of genes and environment and the role they play in behavior ultimately reaches the question of free will. Do we have the freedom to make choices? Does our behavior simply reflect our life experiences? Or do genes dictate our lives? In other words, are we subordinate to our genes? Think about trying to discern between free will and predestination owing to our genes. How could we even do an experiment to find out? Consider that someone does something and then points out that it was done because she or he wanted to do it—that is, because it was a matter of free will. But how do we know that the activity wasn't done because of the individual's genes? Someone offered the following catch-22. If everyone has free will, it must be a hereditary characteristic. If the characteristic is hereditary, then it must have a genetic basis. And if it has a genetic basis, it is not free will. Confused?

In contrast to the notion of strict genetic determinism, there is widespread agreement among geneticists and behavioral scientists that behavior is greatly influenced by culture, societal framework, environmental experiences, and parental influences. Some researchers point out that children of wife beaters sometimes tend themselves to be wife beaters. If true, is this lousy behavior due to heredity or the environment? It could be either or both. Some researchers say that children of alcoholics have an increased chance of being alcoholics, and this may be true in some cases. But just the opposite can occur. Some individuals will absolutely not drink because of the alcoholic environment they experienced while growing up. Regardless of the possibility that they might have genes for the propensity of alcoholism, their environment actually shifted the outcome in their life in the opposite direction.

When discussing the topic of free will, the concept of sociobiology immediately comes into consideration. Sociobiology is a discipline that considers how social behaviors such as aggression, mating, parenthood, love, hate, altruism, and selfishness might be controlled by genes. Presumably, the exercise of these behaviors should improve the chances that the genes responsible would be passed to new generations. This relationship may be possible for mating and parenting behaviors, but certainly less so for other behaviors. The sociobiology discipline is based on the idea of natural selection, and it proposes that even society and culture must ultimately have a genetic basis, and that human progress is related to genes. In a true sociobiological sense, behavioral and emotional characteristics could not be changed by culture, society, or education. The hypotheses surrounding sociobiology have met with much opposition and have generated vehement debate among biologists, sociologists, psychologists, and philosophers, among others. The basic question often debated is whether society and culture emerge from heredity or environment. Data are available to support the role of the environment in human development. Sociobiologists have some data indicating a genetic component at work within some animal societies, but so far there is insufficient evidence for or against sociobiological concepts acting in human populations.

Sociobiology: The New Synthesis, a book written by Edward O. Wilson, a leading authority on the sociobiology of animals, is one of the prevailing treatises on the subject. The book is an excellent discussion on how genetics may influence animal behavior. Wilson's personal research has dealt mostly with ants and their society. Other researchers have studied different insects, fish, birds, and apes, among other species. Some researchers in the past, Wilson included, have registered the possibility that such studies may help to synthesize hypotheses to explain human behavior. In other words, the question has been raised about whether human behavioral and emotional traits are inherited—that is, traits such as altruism, generosity, selfishness, hate, love, anger, tenderness, guilt, fear, pride, ethics, morals, and so on.

Such sociobiological ideas have prompted angry responses among both scientists and the public. The concept has been construed as genetic determinism, and many have felt that the whole notion was errant and that it might justify social inequities. Serious concerns went rampant and people alleged that sociobiology smacks at free will, religion, inequality concerning women, minorities, and even hinders some social programs. Sociobiology was deemed to be a social weapon and up to no good. A few scientists have gone so far as to label sociobiology as another form of racism, white supremacy, sexism, and elitism. You probably understand what is implied by all of this discontent. The woman should stay in the kitchen, barefooted, pregnant, holding a kid on one of her hips while she is making soup for the husband. As another example, turn to the comics in newspapers that often show analogies to genetic issues. Hagar the Horrible was giving the facts of life to his young son. It went something like the following. Women like to wash, clean, sweep, polish, sew, cook, scrub, shop, and gossip. Men like to play games, tell stories, drink beer, eat rich foods, fight, laugh, and sing songs. It was just a comic strip, but to some extent it is the epitome of hard-core sociobiology.

Professor Wilson might have been wise to stop at chapter 26 in his book, leaving out chapter 27. This chapter is the one that brings humans into the realm of sociobiology. In 1978, Wilson was scheduled to deliver a presentation at the American Association for the Advancement of Science Conference in Washington, DC. The talk was scheduled in a fairly large auditorium because the topic of sociobiology had become so controversial. As Professor Wilson began his talk, a vigilante group charged the stage and threw a considerably large container of water on him. The water-throwing event was supposed to be symbolic of someone being "all wet." Wilson has actually modified and softened some of his views on human sociobiology in more recent times. No solid evidence yet exists for the heredity of social attitudes in humans.

The opposition to human sociobiology should not be much of a surprise. Certain sociobiological concepts tend to diminish the importance of cultural background, religious beliefs, free will, and equality of opportunity for women and minorities. On the other hand, it is felt that we should not fear the implications of sociobiology as uncomfortable as they may be. Some geneticists and evolutionists advocate a continued effort to search out the truth with scientific objectivity. They feel that we should not sweep aside the unanswered questions.

Much of the resistance to sociobiological hypotheses about human behavior stems from strong apprehensions about genetic determinism. Such apprehensions are not unwarranted; however, they need to be tempered by the realization that a few behavioral traits are solely genetic. The author has never talked to a geneticist who believed that massive genetic determinism exists. However, data have indicated that some genes have some influence on some behaviors. Still the environment is also believed by most researchers to be very important in the molding of individuals, and not just the environment provided by parents. The environment may be just as important in our life and the choices that we make as are the genes that we own, and possibly more important. Even identical twins with identical genes are not identical in all ways. And just because some genetic influence for a behavioral trait might be real, the trait could still be malleable to a large extent. Behavior is probably due to a vast assembly of nerve cells and many associated molecules. Francis Crick of DNA fame believed that free will was actually located in a certain part of the brain. Neuronal circuitry is extremely complex; hence, predictions of behavior are immensely difficult, regardless of our knowledge about genes and environmental factors. It would simply be great if everyone would just understand right from wrong and abide by it.

FURTHER READING

Baldwin, J. D., and J. I. Baldwin. 1981. *Beyond Sociobiology*. New York: Elsevier.

Barash, D. P. 1977. *Sociobiology and Behavior*. New York: Elsevier.

Burt, R., and R. Trivers. 2006. *Genes in Conflict: The Biology of Selfish Genetic Elements*. Cambridge, MA: Belknap Press of Harvard University Press.

Caplan, A. L., ed. 1978. *The Sociobiology Debate*. New York: Harper & Row.

Dawkins, R. 1976. *The Selfish Gene*. New York: Oxford University Press.

Freedman, D. G. 1979. *Human Sociobiology: A Holistic Approach*. New York: Free Press.

Ruse, M. 1979. *Sociobiology: Sense or Nonsense?* Boston: D. Reidel.

Segerstrale, U. 2000. *Defenders of the Truth: The Battle for Science in the Sociobiology Debate and Beyond*. New York: Oxford University Press.

REFERENCES

Crick, F. 1994. *The Astonishing Hypothesis: The Scientific Search for the Soul*. New York: Charles Scribner's Sons.

Dawkins, R. 1982. "The Myth of Genetic Determinism." *New Scientist*, January: 27–30.

Wilson, E. O. 1975. *Sociobiology: The New Synthesis*. Cambridge, MA: Harvard University Press.

CHAPTER 28

CANCER—
INNOCENT CELLS GONE
ASTRAY

Anytime cells proliferate, things can go wrong, such as gene mutations, chromosome aberrations, and erroneous chromosome numbers. Beyond these events, what else could possibly go wrong within our cells? Well, there is still cancer. Cancer is a disease of the cell like most diseases. A cancer, or neoplasm, develops when for some reason a cell stops listening to the intracellular and extracellular signals and messages that normally regulate growth and development. Simply stated, cells begin to divide out of control, often forming a mass called a tumor. If a tumor lacks the ability to spread in the body, it is benign (noncancerous). If the cancer spreads to other parts of the body (metastasizing), the cancer is described as malignant. Malignant cancers can be fatal even when aggressively treated. Over two hundred distinct types of cancer are known, and almost the same number of causes exist. Together they account for about 20 percent of the deaths in the United States. In 1971, the then president Richard Nixon declared war on cancer, but successes came slowly. Cancer has indeed presented biologists with one of their most formidable challenges.

The life history of cells, called the cell cycle, plays an important role in genetic health. Cancer is an abnormality in the controls of the cell cycle. If the proliferation of new cells approximates the processes of cell death, no significant change in cell number occurs, and the situation is called homeostasis. If cell death greatly exceeds cell proliferation, one of several diseases can develop. Diseases such as Alzheimer's and Parkinson's may be examples of this latter situation. If, on the other hand, cell proliferation greatly exceeds cell death, the abnormal buildup of cells may constitute a cancer. These cellular relationships are the reasons why so many geneticists, cell biologists, and molecular biologists are specifically investigating the cell cycle.

One interesting question is whether cancers are environmental diseases or genetic diseases. The answer is probably yes and yes. Many cancers have been shown to be caused by mutant genes or chromosomal aberrations occurring in body cells (called somatic mutations). The actual mechanisms, however, may be more complex. A cause-and-effect issue also exists here. Do aberrant chromosomes cause cancer, or does cancer cause aberrant chromosomes? Recent studies are showing it is more likely that aberrant chromosomes cause the cancer. In this sense, it means that the disease is genetic. But on the other hand, many cancers do not occur until induced by some factor in the environment. These perturbing factors are called carcinogens and include certain chemicals, radiation, ultraviolet light, and even a few viruses. At the cellular level, cancer is certainly genetic. When a cell, which is committed to be cancerous, undergoes cell division, both of its descendants are also cancer cells, indicating that genetic changes must have occurred within the cell. Usually, the genetic information in our cells is stable. During normal cell divisions, the genetic information replicates over and over again. However, if the genetic information is damaged, the cell's normal fidelity is broken, and the cell may become cancerous. From one progenitor cell, a population of cells will result that has lost its genetic integrity. The resultant cancer cells proliferate and will often invade normal tissues. Without any doubt, at the cellular level cancer has a genetic basis.

A different question is whether individuals can have a genetic predisposition to cancer regardless of their environment. Scientists know little about the inherited risk of cancer. This question is not easy to answer, although hypotheses exist. Is the onset of cancer in individuals within the same family a matter of genetic predisposition or just bad luck? Cancer is indeed sometimes a family matter, but is this due to their sharing of the same genes or their sharing of the same environment? One hypothesis describes the onset of some cancers, or the resistance to them, to some extent as being due to an individual's repertoire of particular genes. If correct, someone with a very low resistance to the onset of cancer could conceivably contract the disease in spite of living in an environment relatively free of carcinogens. In contrast, someone with a high level of resistance might smoke heavily, drink a lot, and generally live a terribly unhealthy life before succumbing to an accident at ninety-six years of age. Everyone probably knows someone who fits this latter scenario. Place no weight on this observation for guidance in your own life, however, because you have little or no information about the array of genes making up your cancer defenses. You do not know much about who you really are in this regard.

Several kinds of genes are good candidates for defense against cancer. Three classes of genes making up the best candidates are (1) genes for the repair of DNA; (2) genes for tumor suppression; and (3) proto-oncogenes. We all have genes for repair of damage to DNA, and the effectiveness of these genes can greatly vary among people. One idea is that when you cease to repair your DNA, cancer occurs. Tumor-suppressor genes do exactly that task. They suppress the formation of tumors. A number of tumor-suppressor genes are now known. Most of them play a role in running the cell cycle in a regulated, normal way.

One of the best-known tumor-suppressor genes is called p53. This interesting gene is multifunctional, but it has a central involvement in the regulation of apoptosis—that is, genetically programmed cell death. *Apoptosis* is a Greek word that means "dropping off." Here is how geneticists think the gene works. Genetically damaged cells, not repaired by our repair mechanisms, can become cancer cells and begin the uncontrolled cell proliferation leading to a tumor. Before this happens, we would like the damaged cell to literally commit suicide. The p53 gene plays an important role in getting the damaged cell to die. But if the p53 gene itself mutates so it is no longer functional, the gene will not cause the damaged cell to die. The cell can then proliferate without p53 restraint. Since cell suicide cannot come to the rescue, the damaged cell can proliferate out of control and become a neoplasm (figure 28-1). How frequently does this happen? Approximately 50 to 60 percent of all human tumors tested have been shown to have a mutated p53 gene in the cells making up the tumor.

A third group of genes implicated in carcinogenesis (origin of cancer) is actually composed of normal genes, at least initially.

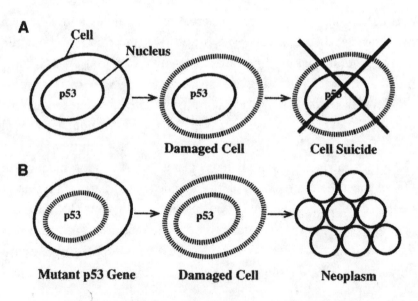

Figure 28-1. The suicidal function of the p53 gene. (A) With a functional p53 gene, a damaged cell will be programmed to undergo cell death, thus preventing its chances to become cancerous. (B) A nonfunctional p53 gene cannot induce cell death in a damaged cell; hence, it can possibly proliferate into a neoplasm.

Among our inventory of about twenty thousand to twenty-five thousand genes, there is a group that has the potential to turn on you. Everyone has them. These genes are called proto-oncogenes. They have roles in normal processes like cell division, cell growth, and development. But mutation can transform proto-oncogenes, and their expression, to cause the normal cell to become a tumor cell. When this transformation takes place, the proto-oncogene is then called an oncogene (cancer-causing gene).

A couple of examples will demonstrate the proto-oncogene to oncogene to inappropriate expression relationship in humans. The *c-myc* gene is located on chromosome number 8 in the human. When this proto-oncogene gets relocated by a translocation to another chromosome (usually chromosomes 2 or 14), the normal gene becomes an oncogene and instigates a cancer called Burkitt lymphoma. This situation is a clear example of the phenomenon of position effect; that is, a gene in different locations within the genome can sometimes express very differently. Other changes in this normal gene can cause a cancer of the lung, cervix, or breast. Another example of position effect is when the *c-abl* proto-oncogene normally found on chromosome number 9 in the human is positioned next to the *bcr* gene on chromosome 22, again by a translocation event. This 9-22 translocation chromosome has become known as the Philadelphia chromosome. The *c-abl* gene is believed to code for a normal enzyme and that's good. The *bcr* gene encodes for a molecule that activates other normal reactions and that's also good. When *c-abl* and *bcr* are in juxtaposition to each other, myelogenous leukemia will occur in the individual—almost always. That, of course, is not good. Two proto-oncogenes combine to become a kind of hybrid gene and definitely oncogenic. The *c-abl* and *bcr* gene combination probably encodes for a different growth factor, causing the leukemia condition. Again, a fundamental important factor for good genetic health is not only having good genes but also an appropriate gene location. Not all proto-oncogenes become oncogenes as a result of chromosome translocations. The *ras* proto-oncogene normally encodes for a regulatory protein. The gene becomes an onco-gene by a mutation that sort of puts it into cruise control, causing a tumor. Many other examples of proto-oncogene-oncogene relation-ships are now known. We all have these proto-oncogenes, and we can only hope that they remain as proto-oncogenes.

A discussion of the *Salmonella*/microsome assay (Ames test) is pertinent to this discussion because the test so explicitly shows the relationship between mutation and cancer. This test is a rather cre-ative system developed by B. N. Ames and his colleagues for the detection of chemical mutagens. One version of the test uses *Salmo-nella* bacteria, a good choice because of the large numbers used per test, as well as its short generation time, low expense, and good sen-sitivity. The *Salmonella* strain has several useful mutations. One muta-tion blocks the bacterium's ability to make a normal coat on its sur-face, thus allowing the test chemicals to more readily penetrate and enter the organism. Another mutation blocks the bacterium's DNA-repair mechanisms, thereby allowing any genetic damage effected by the test chemical to be detected. A third mutation prevents the organism from making its own histidine (called *his–*), an essential amino acid. As a result, the "*his–*" organism cannot grow on an artifi-cial medium lacking histidine. When these bacteria are placed on a histidine-free selective medium, only *his+* (histidine synthesizing) bacteria will be able to survive. In other words, a mutation (reverse mutation) from *his–* to *his+* has to occur for survival of the bacteria. Should a chemical being tested actually be a mutagenic substance, it may change some *his–* bacteria into *his+* bacteria that will be able to grow on the medium. The number of *his+* mutants, easily counted after a couple of days of growing into colonies on the selective medium, is a neat measure of mutagenicity. This system became a favorite piece of scientific testing for the detection of mutagens.

The chemical being tested and an extract from rat liver are mixed with the *Salmonella* bacteria and spread on a Petri dish (plate) containing the selective medium. The rat liver extract (called micro-somal fraction) contains many soluble enzymes and other chemical

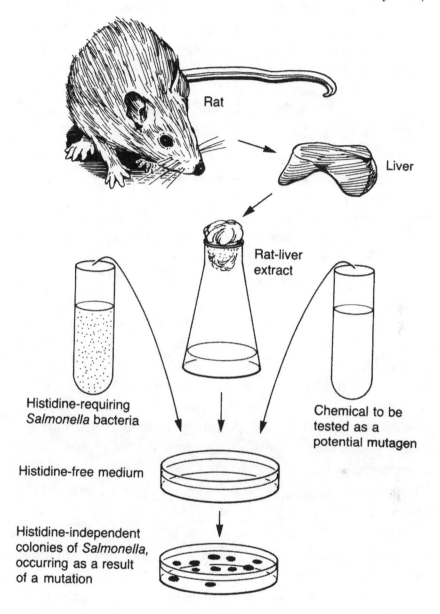

Rat

Liver

Rat-liver
extract

Histidine-requiring
Salmonella bacteria

Chemical to be
tested as a
potential mutagen

Histidine-free medium

Histidine-independent
colonies of *Salmonella*,
occurring as a result
of a mutation

Figure 28-2. The principal components and procedure of the Ames
Salmonella-mammalian liver test for discerning chemical mutagens.

substances from liver cells. The reason for adding this fraction to the test is to simulate the processing of the test chemical as it would normally be processed in mammalian metabolism. Some chemicals may not be mutagenic until they undergo biochemical modification and metabolic activation by the liver. On the other hand, some mutagenic substances may be broken down into harmless substances by the liver. See figure 28-2 for a summary of the *Salmonella* assay.

The *Salmonella* assay has been used to test thousands of chemicals. The assay is deemed to be very sensitive—sensitive enough to detect smokers from nonsmokers by using their urine as the test substance. Smokers have very mutagenic urine. About twenty to forty nasty chemicals can be found in the urine of smokers. Also, it has been shown that extracts obtained from the feces of vegetarians are less mutagenic than those obtained from meat eaters. Some research can be rather unappetizing. In another study, beer was found not to be mutagenic. A beer association funded this study. What a surprise!

A take-home message has been derived from all of these assays. When known carcinogens are used as the test chemicals, approximately 90 percent or more of them were also shown to be mutagenic. Additional research has given us good reason to believe that the reverse is also true; that is, most mutagenic substances are also carcinogens. These latter studies, of course, were conducted on mice, rats, rabbits, monkeys, and the like. The bottom line is that an exceptionally high correlation exists between mutagenesis and carcinogenesis.

Treatment of cancer continues to dramatically improve and even cures are increasing. But will we ever be able to prevent cancer by a vaccine or some other means? A vaccine would probably have to be something that would target the immune system and cause it to recognize and destroy malignant cells. Presently, the best prevention of cancer is to stay out of the way of known mutagens and carcinogens. In other words, don't rough up your chromosomes. But pessimism always exists because the disease is so genetic at the cellular level.

FURTHER READING

Bishop, J. M. 1982. "Oncogenes." *Scientific American* 246: 80–92.

REFERENCES

Ames, B. N., W. E. Durston, E. Yamasaki, and F. D. Lee. 1973. "Carcinogens Are Mutagens: A Simple Test System Combining Liver Homogenates for Activation and Bacteria for Detection." *Proceedings of the National Academy of Sciences* 70: 2281–85.

Ames, B. N., F. D. Lee, and W. E. Durston. 1973. "An Improved Bacterial Test System for the Detection and Classification of Mutagens and Carcinogens." *Proceedings of the National Academy of Sciences* 70: 782–86.

Ames, B. N., J. McCann, and E. Yamasaki. 1975. "Methods for Detecting Carcinogens and Mutagens with the *Salmonella*/Mammalian-Microsome Mutagenicity Test." *Mutation Research* 31: 347–64.

Devoret, R. 1979. "Bacterial Tests for Potential Carcinogens." *Scientific American* 241: 40–49.

Kuhnlein, U., D. Bergstrom, and H. Kuhnlein. 1981. "Mutagens in Feces from Vegetarians and Non-Vegetarians." *Mutation Research* 85: 1–12.

CHAPTER 29

HUMAN GENOME PROJECT

Deoxyribonucleic acid (DNA) is a molecule that is supposedly a household word. A lot of people know something about this fantastic molecule. DNA makes up the genes that make up the chromosomes that make up much of the nucleus found in most of the 70 trillion cells that make up your body. The discovery of DNA, the determination that DNA is *the* genetic material, and the elucidation of the molecule's structure constitute one of the most influential series of events in all of biological history. DNA truly holds the secret of life.

A Swiss chemist by the name of Friedrich Miescher (1844–1895) was interested in the nuclei of cells and went about isolating one of the primary substances found in them. He called the material nuclein, which was actually a crude extract of DNA. Miescher went around collecting puss from bandages in hospitals for his cellular material. Puss is mostly dead white blood cells that have nice, large nuclei, probably making the material a good source for his nuclein. Miescher had no idea that the substance he isolated was the genetic material of life. Eventually the substance was better purified by biochemists, and its composition was determined. The substance was found to be composed of a phosphate group, a type of sugar (deoxyribose), and four different nitrogenous bases. The nitrogenous bases are called guanine, cytosine, thymine, and adenine, simply referred to as G, C, T, and A, respectively.

To better understand what the Human Genome Project accomplished, one needs to take a closer look at the structure of the DNA molecule. When one of the nitrogenous bases is attached to a deoxyribose sugar at one particular point, and a phosphate group is attached to the sugar at another particular point, the resultant chemical group is called a nucleotide (figure 29-1). An entire sequence of these nucleotides chemically bound together by way of the phosphate units forms a long chain, literally thousands of the C, G, T, and A nucleotides joined together in a very specific order. The order of the nucleotides is the very essence of what genes are all about. This long chain of nucleotides is now called a polynucleotide strand or chain (figure 29-2). Two polynucleotide strands are positioned side by side, each G combined with a C and each A combined with a T. These combinations are known as complementary pairings. Molecules take up space and have shape just like all matter. Therefore, the nucleotides in one of the two polynucleotide strands need to be inverted in order to have a good fit between the C and G and the T and A. The molecule with this arrangement is described as being antiparallel, as opposed to being parallel (figure 29-3). Finally, this long ladderlike molecule twists into a helix, like a spiral staircase. Since the molecule has two sides, it is called a double helix (figure 29-4).

Again, the fit between G and C and A and T is described as being a complementary fit; consequently, each combination makes up a base pair (bp). The DNA found in every cell of the human body is considered to have a length of about 3.3 billion of these base pairs. And this entire sequence of base pairs in the human genome has been determined. So how does your bp sequence differ from your neighbor's? Not very much! Maybe a nucleotide here and there, but

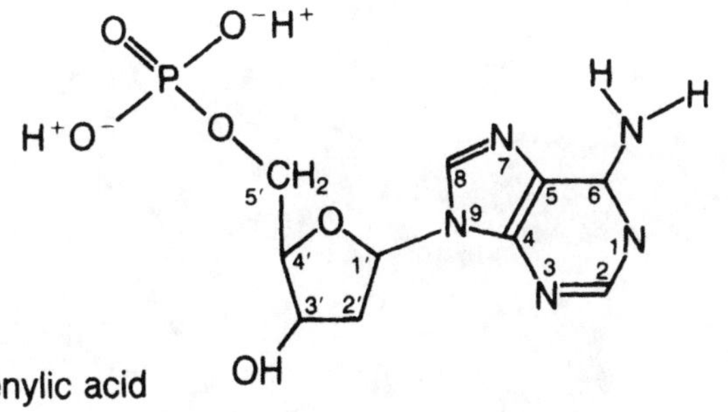

Deoxyadenylic acid

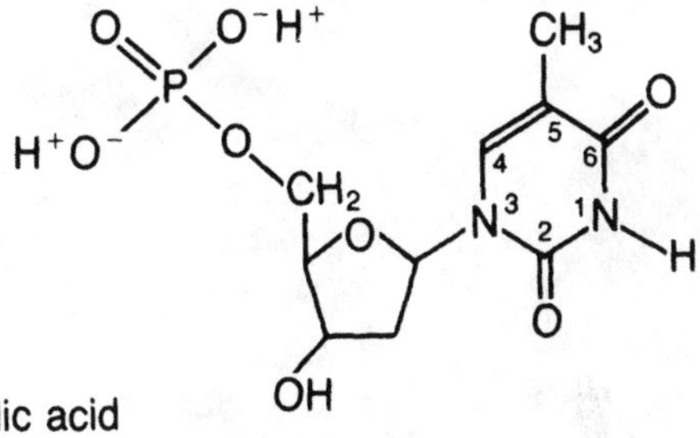

Deoxythymidylic acid

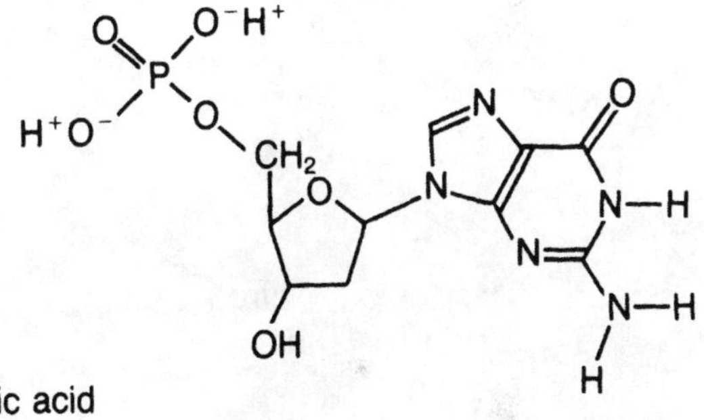

Deoxyguanylic acid

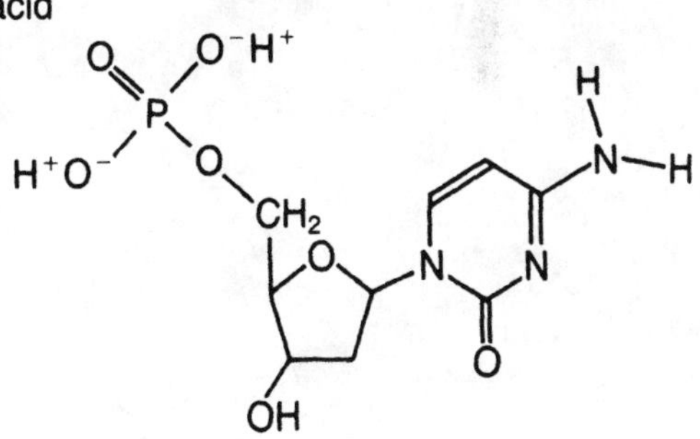

Deoxycytidylic acid

Figure 29-1. The four common nucleotides found in DNA, each consisting of a deoxyribose sugar molecule, phosphoric acid, and one of the four nitrogenous bases.

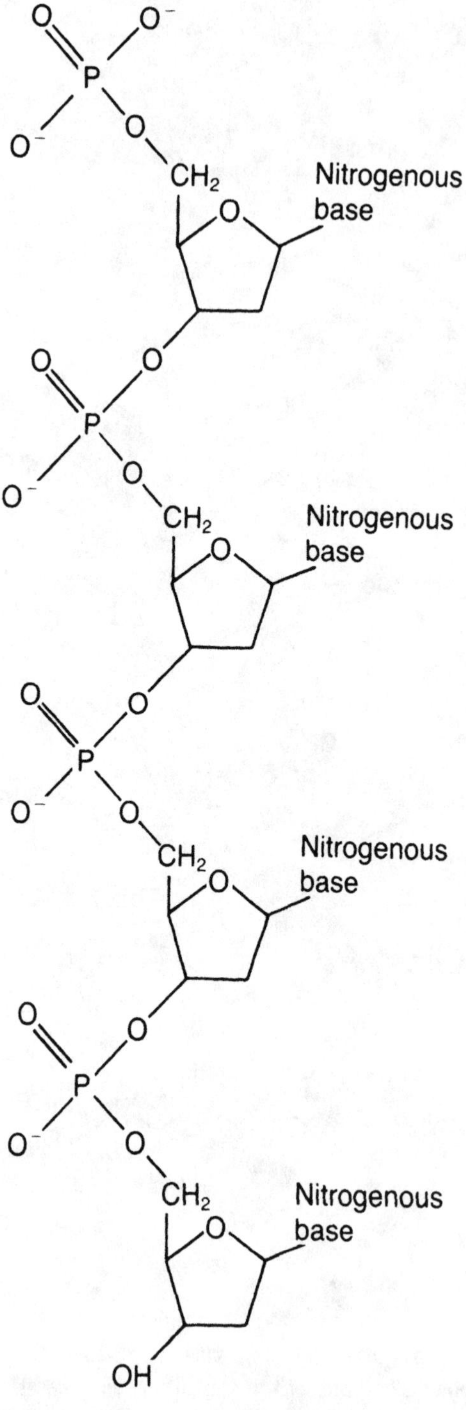

Figure 29-2. A polynucleotide strand consisting of nucleotide subunits. Although only four nucleotides are shown, such strands actually consist of many thousands of nucleotides.

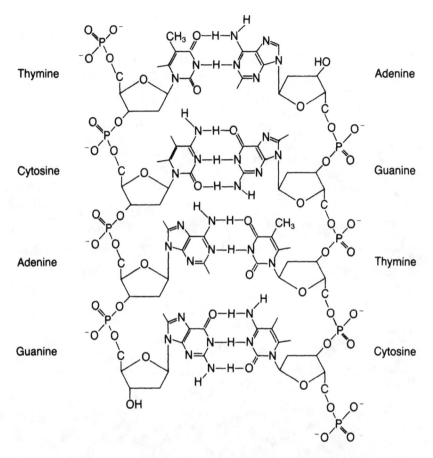

Figure 29-3. A double-stranded segment of DNA showing four nucleotide pairs. Note the complementary pairing and the antiparallel arrangement of the two polynucleotide strands. The chemical designations refer to the nitrogenous base found in each of the nucleotides.

on the other hand, small differences in the sequence can indeed be the cause of much phenotypic variation (appearance) within our species. Your bp sequence would differ slightly more from that of a chimpanzee than from your neighbor, a little more from a rat, more from a toad, and still more from bread mold. These differences are the result of evolution. However, much of the DNA sequence would be the same or very similar among all of these different organisms. Scientists refer to the nearly identical sequences among organisms as

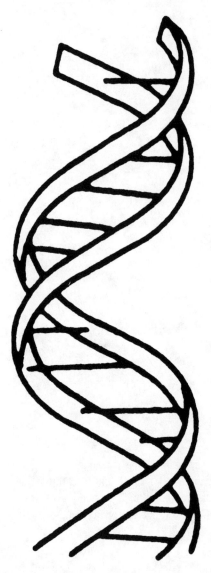

being "conserved." Such sequences are usually those that are essential in keeping the organisms alive, regardless of whether you are a human, a chimpanzee, a rat, a toad, or bread mold.

Ingenious methods have been developed to determine the exact sequence of all those A, T, C, and G nucleotides in the genome (DNA of an organism). The number of bps range from a couple million in bacteria to 3.3 billion in humans. The result of this sequencing is a complete genetic blueprint for that particular organism. The complete sequence has been accomplished for a number of organisms, including a bacterial species, the fruit fly, a small mustard weed, a little roundworm, yeast, rice, our closest relative, the chimpanzee, and of course, the human. Some crop plants are also being sequenced. Knowing the sequence of genomes allows scientists to isolate genes that control specific functions, which, in turn, will lead to useful information in

Figure 29-4. The DNA molecule shown as its usual double helix structure.

all sorts of biological endeavors. The Human Genome Project was a massive undertaking believed to have tremendous potential, especially in medical and other biological research.

In the early 1970s, it took two years to sequence twenty base pairs of DNA. By working at this pace without any improvement in the technique, it would take a period of 310 million years in order to complete the entire human genome. The project just would not have happened. However, new automated methods came about, allowing researchers to sequence thousands of bps per day, and with a number of laboratories simultaneously working on the project, the entire genome was completed in just a few years. The project cost about three billion dollars. The complete genome sequence will fit on a hard disk of a desktop computer. One initial finding is that much of our DNA is not made up of functional genes. Probably about 98.5 percent of the human genome does not code for any kind of message. We now know that the human genome has about twenty thousand to twenty-five thousand actively expressing genes. This gene number is not much different from some of the other multicellular organisms.

Once again, the nucleotide base pairs of A–T and G–C make up the genetic language for all of life. The study of this genetic language is called genomics. This genetic endeavor promises to be a dominating pursuit among geneticists during this century. Geneticists will probably not find genes for nagging, adventure, gambling, giggling, argumentative nature, or road rage. However, researchers hope to increase our knowledge about mutated genes that cause at least several thousand diseases. Knowing the human genome gives them a new set of tools. Not only do medical scientists plan to identify medical diseases, but they may possibly predict vulnerability to such diseases, and even susceptibility to infectious diseases. And new treatments may emerge from these studies. Hopes are frequently expressed that the information will rapidly lead to new treatments against cancer. Gene therapy promises to become more effective and reliable. It might also be possible to eventually monitor genetic damage by various agents in our environment. The information from DNA sequencing will have impacts on drug discovery. Drug treatment could even become individualized—that is, drugs designed to fit a patient's genetic profile. Scientists already know that drugs behave differently in different individuals. People may eventually have personalized medicine. Still other rich biological secrets might be buried in the human genome. The Human Genome Project should provide an avalanche of information and application.

Not all is glee, however, because of possible misuses of the information. Some people fear a threat of genetic discrimination by health insurers and employers, and that the information could create a health bias. People are concerned that someday our genetic code will be imprinted on an identification card that, in some cases, could be used to the individual's disadvantage. Others feel that the danger is grossly overblown. Genetic advances are indeed often controversial.

In still another area of study, many of the questions concerning the specific makeup of a human may be answered by the information obtained through DNA sequencing. Evolutionists believe that humans and chimpanzees had a common ancestor, with the split from each other occurring about five to seven million years ago. Our closest relative appears to be the chimpanzee, but the gorilla and other primates are also fairly close cousins. By outward appearances, marked differences are quite apparent. We are more bipedal (walk on two feet) and have a different gait, an advanced cognitive function, and a complex language. The African great apes (chimpanzees and gorillas) also have a lot more hair than most of us, although one might find some exceptions. Nonetheless, we can usually distinguish humans from apes.

Much focus has been placed on the differences between humans and the African great apes. We need to go beneath the hairiness and discuss chromosomes, genes, and DNA. Chimpanzees, gorillas, and orangutans have twenty-four pairs of chromosomes (forty-eight); all other great apes have twenty-three pairs of chromosomes (forty-six), just like humans. From these chromosome studies it appears that two chromosomes have been fused together in some primates and not in others. These results are made possible from chromosome-banding techniques that display specific patterns for each chromosome of different species (figure 29-5). The chromosome banding patterns are

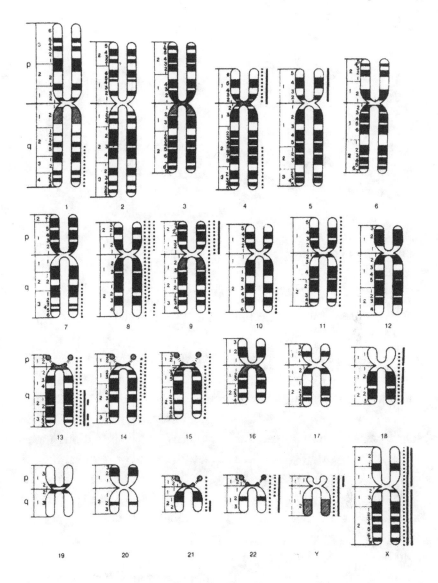

Figure 29-5. One type of banding technique illustrating the differences among the twenty-three pairs of human chromosomes. From Paris Conference, 1971: *Standardization in Human Cytogenetics*. Birth Defects: Original Article Series 7 (New York: National Foundation, 1972). Courtesy of the National Foundation.

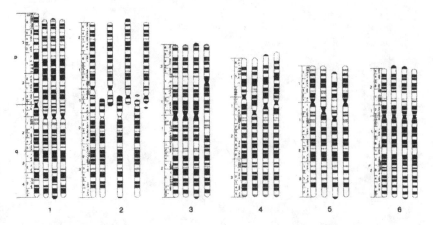

Figure 29-6. Comparison of the banding in six chromosomes (1 through 6) in humans and other primates. From left to right: human, chimpanzee, gorilla, and orangutan. From J. J. Yunis and O. Prakash, "The Origin of Man: A Chromosomal Pictorial Legacy," *Science* 215 (1982): 1525–30. Copyright © 1982 by the American Association for the Advancement of Science. Reprinted with permission of the publisher.

very similar among all primates. You could hardly tell a human from an ape by simply observing the chromosomes (figure 29-6). One would have to check out outward appearances.

Next, consider DNA and genes. Some studies show a genetic difference between humans and chimpanzees to be only about 1 percent; however, others feel this slight difference is questionable. Nonetheless, differences at this level between humans and our closest relative, the chimpanzee, are believed to be due to changes in relatively few genes. This difference calculates to only 1 to 2 percent of the genes that make us humans and not the other primates, which would be about a total of three hundred to five hundred genes. And remember, humans and probably other primates have a total of approximately twenty thousand to twenty-five thousand genes.

So a kind of enigma emerges. If we are so similar at the DNA level, why are we so different at the phenotypic (appearance) level? The whole story is not yet known, but a few hypotheses have been offered. One major difference between humans and other primates continually stands out—that is, the unique human brain. Phenotypically, humans are distinctly different from apes with regard to cognitive function. The human brain shows differences from other primates in size, organization, and function. Most gene differences between humans and apes seem to be at the level of the brain and its development. One study compared 20,000 pieces of human DNA with that of chimpanzees and rhesus monkeys with regard to gene expression—that is, the action of genes. Only 165 genes showed a difference in gene expression between humans and the other primates, and most of these 165 genes were those that dealt with the development of brain tissue. Other studies have shown the human brain to churn out more messenger molecules (called RNA) than the chimpanzee brain, even though other tissues like the liver did not show a difference in the production of messenger molecules.

Some researchers also think that much of this brain difference has to do with timing during the developmental period in gestation; that is, a great amount of effect might be dependent upon when genes will be "on," and very important, for how long. Genes going on and off is a phenomenal orchestration, something like a genetic piano. The idea often posed is that some critical genes for brain development, regardless of whether they are similar to the chimpanzee, are expressing more effectively and for longer periods of time during human development. In other words, "human" in a genetic sense may be tied to brain development, and even a small number of genes affecting brain tissue and neuron circuitry could possibly make a big difference. In addition, some of these genes may be master genes—that is, genes that regulate many other genes. The perceived small differences could, therefore, snowball into effects upon hundreds or even thousands of other genes. The timing of genes going on and off may be of utmost importance.

None of these ideas are definitive at this time. Scientists are bound to learn more about what constitutes a human being from the information obtained through the Human Genome Project. In addition, obtaining more DNA sequences from other primates for comparison will certainly be informative. Scientists also need to study these sequences in relation to the biochemical properties of the proteins for which they code, and in relation to the physiology of the organism. A difference of three hundred to five hundred critical genes between us and the hairy ones could be very substantial at the level of function.

FURTHER READING

Bishop, J. E., and M. Waldholz. 1990. *Genome.* New York: Touchstone, an imprint of Simon & Schuster.

Gazzaniga, M. 2008. *Human: The Science behind What Makes Us Unique.* New York: Ecco.

Gibson, G., and S. V. Muse. 2009. *A Primer of Genome Science.* Sunderland, MA: Sinauer Associates.

Hawley, R. S., and C. A. Mori. 1999. *The Human Genome.* New York: Academic Press.

Marks, J. 2002. *What It Means to Be 98% Chimpanzee: Apes, People, and Their Genes.* Berkeley: University of California Press.

McConkey, E. H. 2004. *How the Human Genome Works.* Sudbury, MA: Jones and Bartlett.

Murphy, T. F., and M. A. Lappe, eds. 1994. *Justice and the Human Genome Project.* Berkeley: University of California Press.

Primrose, S. B., and R. M. Twyman. 2004. *Genomics: Applications in Human Biology.* Malden, MA: Blackwell.

Ridley, M. 1999. *Genome: The Autobiography of a Species in 23 Chapters.* New York: HarperCollins.

REFERENCES

Crick, F. H. C. 1957. "Nucleic Acids." *Scientific American* 197: 188–200.

Miescher, F. 1871. "On the Chemical Composition of Pus Cells." In *Great Experiments in Biology*, edited by M. L. Gabriel and S. Fogel, 233–39. Englewood Cliffs, NJ: Prentice-Hall, 1955.

Paris Conference. 1971. *Standardization in Human Cytogenetics.* Birth Defects: Original Article Series 7 (1972). New York: National Foundation.

Watson, J. D. 1968. *The Double Helix.* New York: Atheneum.

Watson, J. D., and F. H. C. Crick. 1953. "Molecular Structure of Nucleic Acids." *Nature* 171: 737–38.

Yunis, J. J., and O. Prakash. 1982. "The Origin of Man: A Chromosomal Pictorial Legacy." *Science* 215: 1525–30.

CHAPTER 30

SKIN COLOR

Not all human genetic traits follow a simple Mendelian mode of inheritance. Many of them are not either/or situations. Instead, some traits are continuous and quantitative meaning that their level of expression or magnitude can fall anywhere within a certain range. The best explanation for this variability in expression is polygenic inheritance (many genes) combined with a strong environmental influence. Skin color is fairly well documented as being a polygenic trait. Three to five gene pairs (six to ten genes) probably determine whether you are one of those very light persons at one end of the color spectrum, or a very dark person at the other end, or anything in between the two extremes.

Polygenic inheritance is more complicated than simple Mendelian inheritance because it involves more than one gene pair. Each gene pair behaves in Mendelian fashion with regard to its transmission to the next generation, but usually the effect of each gene is small and additive. The individual effects of the genes add together to produce a particular quantitative level of expression for the characteristic. The effects of the various genes are usually assumed to be equal, but this assumption may not always hold true for some traits, since certain genes may have greater effects than others. In addition, truly additive genes do not show dominant and recessive relationships. The slight effects exerted by various alleles are cumulative, that is, summed to relate to a particular expression.

Consider a simple hypothetical situation involving three gene pairs. Each uppercase-designated allele exerts a certain enhancement effect on a phenotype above a minimum; lowercase-designated alleles do not. An AABBCC × aabbcc cross (F_1) would result in

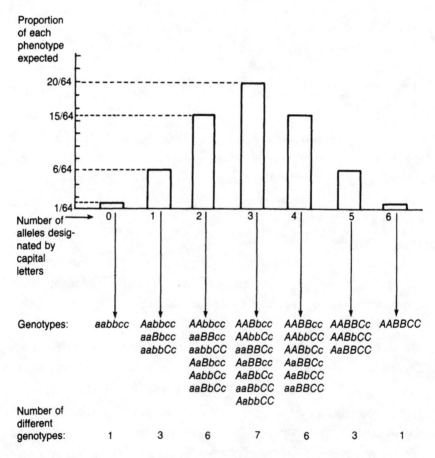

Figure 30-1. The proportion of different F_2 progeny expected from the hypothetical cross AaBbCc × AaBbCc, assuming that three pairs of genes are involved in the expression of a polygenic trait.

AaBbCc progeny. The F_2 progeny of an AaBbCc × AaBbCc cross, however, would show twenty-seven different genotypes and seven different phenotypes, as shown in figure 30-1. The most common phenotype is expected to occur about twenty times out of sixty-four progeny, and the same phenotype is produced by seven distinct genotypes, all of which have three uppercase-designated alleles.

In this basic example, we are assuming that all of the genes contribute equally to the phenotype, and that the genes express in a completely additive way. Again, this may not always be the case. Furthermore, the environment usually plays a role in determining the phenotype, sometimes a very large role. The expression of certain traits in a group of genetically identical individuals can vary considerably as a result of environmental differences.

In figure 30-1, three gene pairs were able to produce seven different phenotypes. With more gene pairs, the number of phenotypic classes would increase, and the differences among some of them would lessen. If the environment also affects the trait, closely similar phenotypes would tend to overlap. At this point, distinct classes of the phenotype would no longer be recognizable. The distribution of phenotypic expressions becomes a continuous curve (a curve of normal distribution), and it becomes difficult to distinguish hereditary from environmental effects.

The heredity of skin color in humans has always been of much interest. This discussion of the heredity of skin color is not referring to whether you are an albino versus a melanin-synthesizing individual; rather, this analysis refers to the extent to which you make melanin, assuming you do not have albinism. Albinos cannot synthesize any melanin. Melanocytes are cells in the skin that distribute pigment granules throughout the skin layers. Actually, light and dark people have the same amount of melanocytes, but the variation in skin color is due to the number and distribution of the melanin pigment in the upper layers of the skin. Of course, exposure to sunlight can also increase the melanin synthesis and distribution.

People, therefore, can be extremely fair skinned on one hand and very dark skinned on the other hand, and every gradation of skin shades between the two extremes. So the question has always been about how many gene pairs are playing a role in this gradation of melanin synthesis observed in the skin and hair of humans? As early as 1913, C. B. Davenport concluded that only two gene pairs regulated human skin coloration. Although his methods were crude (very crude) by today's standards, they did reflect some ingenuity in his attempt to quantify the hereditary cause of different skin colors. Davenport studied the progeny of marriages between whites and blacks in Jamaica and Bermuda. In those days, these islands were about the only places that he could study mixed marriages and obtain that kind of data. He used paper disks with colored sectors of varying sizes to measure the shade of the skin of each experimental subject. He literally spun the disk like a top on a central axis so that the different shades would appear to blend into one color. The resulting whirling of the disk could be made to match the individual's skin color simply by changing the proportions of the colors in the disk, especially the black sector. The proportion of the black sector in the top was recorded as the measure (an actual number) of skin pigmentation. Consequently, Davenport was able to study the trait as a quantitative one. The crude measurement was taken on or under the upper arm, an area of the body that Davenport assumed to be usually protected from sunlight.

Davenport's research was a strange way to collect data—that is, by spinning little cardboard tops on the body of each person in his study. At any rate, the information was useful in gaining a better understanding of polygenic inheritance. The F_1 children from white × black matings (the two extreme skin colors) showed a skin color closely intermediate to their parents. This result was the first good indication that the trait was due to polygenic inheritance. It is almost like considering simple arithmetic because a high measurement plus a low measurement divided by two equals a measurement intermediate of the other two measurements. When Davenport checked the F_2 children from parents who were both intermediates with his spinning top, he recorded that about one in sixteen of the children was very dark and one in sixteen was very light, prompting him to conclude that only two gene pairs were involved in the shade of our skin. If we assumed this conclusion to be true, then AABB would be very dark, aabb would be very light, and AaBb individuals would be intermediate. When an AaBb individual mated with another AaBb individual, the result would be progeny whereby 1/16 would be expected to be AABB (very dark) and 1/16 would be expected to be aabb (very light). These proportions are what Davenport thought he observed.

Davenport's reasoning made good genetic sense, but it was the wrong answer. His methods were just too crude, and the samples were just too small. In addition, the environment (sunshine) probably played a larger role in the skin color than Davenport presumed, regardless of his attempt to remove sunshine as a factor. Skin color is certainly a multifactorial characteristic; that is, its variation is due to polygenes and the environment. Some studies have shown the variation of skin color due to hereditary causes to be about 65 percent, and that due to the environment is about 35 percent.

Davenport's investigation is generally regarded as a creative piece of work, and the basic conclusion that multiple genes (polygenes) control human skin coloration was correct. Modern analysis has used sophisticated instrumentation called reflective spectrophotometry. The instrument measures the amount of light reflected from a surface—in this case, the skin. The darker the skin, the less light that will be reflected. These methods are much more quantitative than spinning cardboard tops on somebody's skin. Hence, most scientists today believe that skin color is due to more like three to four or even five to six gene pairs. And some genes may be more important than others. Certainly the skin color trait is complex, and its heredity is not definitively established. It may not be the simple genetic model as described.

Assume that five gene pairs are involved in the intensity of the trait, and that the gene pairs act in an additive way. Then, ten genes and eleven phenotypes would be possible. In other words, the number of genes for melanin enhancement in the skin could be ten, nine, eight, seven, six, five, four, three, two, one, or zero, with ten being very dark and zero being very light. Superimposing the strong influence of the environment on the expression due to genes leads to an understanding of why there is such a continuous gradation of skin colors in the human population.

Regardless of what one sometimes hears, a very black person mating with a very black person can only result in black children. A very white person mating with a very white person can only result in white children. A black-with-white mating will result in children having an intermediate skin color. An intermediate-with-intermediate mating has the possibility of yielding children all the way from very black to very white and every shade of skin in between. It is a matter of the genes for the enhancement of melanin, and how they segregate to the progeny.

REFERENCES

Davenport, C. B. 1913. "Heredity of Skin Color in Negro-White Crosses." *Carnegie Institution of Washington* 188: 1–106.

———. 1926. "The Skin Colors of the Races of Mankind." *Natural History* 26: 44–49.

CHAPTER 31

GENE POOLS AND RACE

The total genetic information making up the members of a population of sexually reproducing organisms is known as the gene pool. Thinking in this case has to be turned to the concept of populations of genes (or alleles). In nature, few organisms live alone; rather, they are part of populations. A genetic population refers to any group of interbreeding individuals. Population genetics is the study of the distribution of genes in populations and how gene and genotype frequencies are maintained or changed. Population genetics is also the study of factors involved in evolution. Such population studies are of much interest to ecologists, evolutionists, agricultural scientists, epidemiologists, and medical geneticists, in addition to those in many other scientific disciplines.

Soon after Gregor Mendel's work was discovered, some biologists, even a few who called themselves geneticists, were quite concerned that dominant genes would eventually overtake the recessive genes in the population. They thought that a three-to-one ratio of people with the dominant form to people with the recessive form would be reached in the population. If true, this change would certainly be a serious dilemma. Significant problems are associated with some of the dominant alleles: brachydactyly with very short chubby fingers; polydactyly with too many fingers and/or toes; ichthyosis with a scalelike skin condition; achondroplasia with a type of dwarfism; and Huntington disease with severe neurological disorders. The good news is that this initial notion about heredity was completely erroneous. Three-fourths of the population is not going to be made up of six-fingered dwarfs with skin and neurological problems.

Two publications independently appeared in 1908 with explanations showing why dominant alleles will not increase in the population. A mathematician, G. E. Hardy, wrote a letter to the editor of the *Science* journal that sounded much like a reprimand of the non-mathematical biologists who were making such erroneous claims. Wilhelm Weinberg, a German physician, also pointed out the correct mathematical relationships, although his article appeared in an obscure journal much like Mendel's experiments. Hardy and Weinberg made their case with simple middle school algebra. They showed that allelic frequencies, dominant or recessive genes, would not change from generation to generation unless some force acting upon them perturbed the population. This conclusion became the Hardy-Weinberg law (or principle). The basis of this principle has given rise to the field of population genetics. The Hardy-Weinberg principle is also one of the bases for evolutionary thought.

To understand the rudimental aspects of population genetics, one needs to begin with gene frequencies. Consider the MN blood system in humans. Most people are aware of the ABO blood system and maybe even the RH system, but a number of other blood systems exist. One of these systems is the MN system, which is an example of codominance. Anyone carrying both of the alleles, M and N, will show both of the M and the N antigens on their red blood cells. Such codominant systems are great examples to use for the demonstration of population genetics concepts. Upon testing individuals in a population, researchers will know the exact frequencies of the MM, MN, and NN genotypes in the sample group.

Dominance cannot cloud the situation, and the calculations become simple arithmetic.

For simplicity purposes, assume that a small population of 100 individuals has the following frequencies of MN blood types:

50 persons with MM
20 persons with MN
30 persons with NN

Since everyone has two alleles for this blood system, 200 alleles (100 × 2) is used in the calculations. The frequency of M is 50 MM people, equaling 100 M alleles, plus 20 MN people with one M allele, for a total of 120 M alleles out of the total 200. Therefore, 120/200 equals a frequency of .60, or 60 percent. The frequency of N is 30 people × 2 alleles, equaling 60 N alleles, plus 20 people with one N allele, for a total of 80 N alleles out of 200. Therefore, 80/200 equals a frequency of .40, or 40 percent. So the gene frequencies are:

M = .60 and N = .40

The idea that gene frequencies will not change in a population from generation to generation (the Hardy-Weinberg law) is only true as long as certain factors do not perturb the population. These disturbances are mutation, selection, random drift, and migration. Nonrandom mating can change genotype frequencies—that is, the number of homozygotes versus heterozygotes in a population—but not the gene frequencies. Nonrandom mating includes two distinctly different types. The mating of likes would certainly have an effect on genotype frequencies—for example, talls with talls, shorts with shorts, and so on. Matings of opposites would also constitute a form of nonrandom mating—like talls always mating with shorts, and so forth. Neither of these types of mating can be considered randomness.

Mutations have rightly been designated the ultimate source of all variability. Natural selection, on the other hand, is the driving force determining which mutations will thrive and which ones will be weeded out. Hence, natural selection is a powerful force in effecting changes in gene frequencies in a population. Most mutations, however, usually do not cause rapid changes in the gene frequencies of a population. First, mutations occur infrequently. Typical mutation rates are about one in 10,000 at the high end and about one in 1,000,000,000 (one billion) at the low end. In addition, many mutations can occur in both directions—that is, from a normal allele to a mutant allelic form, or from a mutant allele back to a normal allelic form. So every time that a gene "A" becomes "a," an "a" gene becomes an "A." This back-and-forth change occurring without changing the overall gene frequencies is called equilibrium. Still, when a mutation results in a distinct advantage for an organism in a particular environment in which the organism resides, such a mutation can then play a significant role in evolution over long periods of time because evolution is change.

Genetic drift is also known as random drift. This concept deals with chance events. The founder's principle (effect) is an example of genetic drift. A change in gene frequency can occur when only a few parents are involved in beginning the next gene pool by reproduction (figure 31-1). A good example of genetic drift probably occurred on an island in the south Pacific. The island is inhabited by about 1,600 persons. Strangely, however, 7 percent of the people on this island have a "rare" genetic recessive eye disease. Those affected cannot tolerate bright light. According to Hardy-Weinberg calculations, researchers came up with the following numbers.

	On the island	In the world
Genotype frequency of the eye disorder	.07	.000009
Gene frequency of the eye disorder	.265	.003
Frequency of carriers for the disorder	.39	.0059

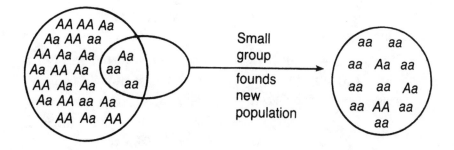

Figure 31-1. An illustration of the founder's principle. The gene frequencies in the new population can be very different from those in the original population.

The disorder is certainly not rare on this island compared to the rest of the world. A possible explanation has been offered for this disparity between the island and the rest of the world. A typhoon supposedly struck the island in the late 1800s. Only about fifteen to thirty people survived, and some of the survivors just happened to have the rare eye disorder. The survivor result could have been due to chance, or quite possibly because the persons with the eye disorder were hiding in caves protecting themselves from sunlight. Regardless, they became a large proportion of the individuals involved in beginning the next population following the typhoon. The allele for this eye defect, therefore, became a significantly high frequency in the next gene pool. This population serves as an excellent example of genetic drift.

Other good examples of genetic drift are apparent. Crigler-Najjar syndrome is a severe liver disease and very rare. Half of all cases of this disease in the United States occur in the religious isolates of the Pennsylvania Amish. It is believed that the propagation of this allele may have begun with one founding family about sixteen generations ago.

Migration is still another way for gene frequencies in a population to undergo changes. One can think of migration as gene flow; that is, genes can flow from one breeding population to another. Many examples of migration are apparent in the world of plants and animals. Animals sometimes migrate from one region to another. Plants migrate through their pollen and seeds being dispersed by wind, water, and animals. Migration has also occurred in human populations. Invading and occupying armies have always been good at creating gene flow. Slavery in the United States is another prime example of migration. A particular gene flow, genes from white men to black women, took place before the boats with the slaves even reached the North American shoreline. White–black matings were perceived to produce progeny who were the best slaves. In fact, it has been estimated that about 30 percent of the genes now found in the overall United States black population are of white-population origin. The change in gene frequency is the result of migration, and it is every bit as real as when a population travels from one continent to another. Gene flow has definitely occurred.

It needs to be pointed out that many geneticists are convinced that race is nothing more than a social and cultural issue, and not a genetic concept. They believe that race has no biological meaning; that only one race exists, and we call it the "human" race. Physical differences have always been used to place human groups into categories called races, such as white, Asian, Hispanic, Native American, African American, and so on. Of course, differences in appearances do exist and contribute to ideas about race. However, these outward differences are deemed to be superficial, and they do not reflect heredity very well at all. But science is not always cut and dried, and not all scientists agree with the notion that a genetic concept of race is meaningless. In fact, some vehemently disagree with this genetic stance.

The genetic perspective in what we call race is beginning to accumulate a vast amount of data. To some extent, this information is coming from the Human Genome Project. This venture was a massive undertaking by scientists to obtain the exact sequence of our base pairs (A, T, G, and C) in DNA, all three billion of them. As one consequence, scientists found that humans are strikingly homogeneous when they get down to the DNA base pair nitty-gritty of things. Remarkably, 99.9 percent of our genes are the same from one individual to another if the 0.1 percent difference is attributed to single nucleotide differences in the DNA. However, differences in gene copy number could increase the overall difference among individuals upward to 0.5 percent. More important, the 0.1 to 0.5 percent differences do not line up by the previously designated races; rather, the differences are randomly scattered throughout the populations of the world. Members of the "racial" groups have just as many genes in common with members of other racial groups as they do within their own groups.

Traits used to classify people into what we have called races are controlled by very few genes. But, of course, we see certain alleles (different gene forms) in higher frequencies within particular population groups than in other groups. The cystic fibrosis allele has a higher frequency among whites than other groups. The Tay-Sachs allele has the highest frequency among Ashkenazi Jewish people. Huntington disease can be found in a relatively higher frequency among the people of Lake Maracaibo, a region of Venezuela. Blacks have a higher frequency of sickle cell anemia than other groups. Blacks also have a higher allele frequency for enhanced melanin production in their skin. Some of these differences are believed to be due to evolution, whereby humans responded to climate and other environmental factors. Hence, such differences may be a matter of ethnicity rather than race. Ethnicity encompasses both genetics and culture. Undoubtedly, some of these different allelic frequencies are due to the founder effect. Again, the founder effect is the initial establishment of populations by a small number of individuals who may not be representative of the allelic frequencies expressed within the larger parent population.

An important point is that different populations of people placed into these racial groups generally do not have *different genes*. They simply have *different frequencies* of the allelic forms of these genes. For example, all people regardless of their racial pigeonhole have either dry earwax or wet earwax. Such genetic data are due to the existence of different alleles for making earwax. The frequency

Figure 31-2. Dennis the Menace © North American Syndicates.
Reprinted with permission.

for the dry crumbly stuff in your ears ranges from only 7 percent in American blacks to a whopping 98 percent in the northern Chinese. Another example is whether you can taste a harmless chemical known as phenylthiocarbamide (PTC). Both of the alleles for determining whether you are a taster or a nontaster of this chemical are found in all racial groups, just in different frequencies. Some people cannot taste the stuff, while others grimace at its bitter taste, regardless of the group to which they belong.

Theoretically, assume that blacks and whites in the United States began to interbreed at random for the next fifty or one hundred generations—that is, marrying and mating without any "racial" hangups. With this much random gene flow over such a long period of time, very few true whites or true blacks would exist in the United States. Skin color and other characteristics normally associated with so-called race would practically blend away. Granted, the average skin color would be closer to that of whites than that of blacks because the country is composed of more whites than blacks. It is somewhat like mixing paints. At any rate, the "races" relative to skin color in this hypothetical example would be nearly obliterated. The concept of different cultures, of course, is still useful and important, and it should be preserved. However, for some geneticists, race in a genetic sense really doesn't exist. Many geneticists and anthropologists think that it is time to bury the whole contrivance of race.

FURTHER READING

Crow, J. F., and M Kimura. 1970. *An Introduction to Population Genetics Theory.* New York: Harper & Row.

Ebling, F. J., ed. 1975. *Racial Variations in Man.* New York: John Wiley & Sons.

Ehrlich, P. R., and S. S. Feldman. 1977. *The Race Bomb: Skin Color, Prejudice and Intelligence.* New York: Quadrangle, the New York Times Book Company.

Goldsby, R. 1971. *Race and Races.* New York: Macmillan.

King, J. C. 1971. *The Biology of Race.* New York: Harcourt Brace Jovanovich.

REFERENCES

Boyd, W. C. 1950. *Genetics and the Races of Man.* Boston: Little, Brown.

Davenport, C. B. 1926. "The Skin Colors of the Races of Mankind." *Natural History* 26: 44–49.

Hardy, G. E. 1908. "Mendelian Proportions in a Mixed Population." *Science* 28: 49–50.

Weinberg, W. 1908. Translated in English in S. H. Boyer, *Papers on Human Genetics.* Englewood Cliffs, NJ: Prentice-Hall, 1963.

CHAPTER 32

FOUNTAIN OF YOUTH

Everyone probably knows how life expectancy is calculated. The life expectancy is somewhere around seventy-eight years for people in the United States. When a child is born, he or she can expect to live about seventy-eight years, the average age of death. Many die younger and many die older. Life expectancy has been extended at an amazing rate. In 1900, the life expectancy was only about forty-six years. Much of the great increase, of course, is due to fewer young people succumbing to various diseases, infection, poor nutrition, and a lack of safety measures. But life span is a very different benchmark. The maximum life span, on the other hand, is about 115 to 120 years, albeit a very few rare exceptions have occurred whereby a person might live for an even longer time. Most people do not reach this maximum life span because of disease, accidents, and other misfortunes. It is interesting to note that modern medicine has greatly increased the life expectancy but not the life span. The upper limits for the life span have hardly changed at all since the beginning of such records. This observation is the first clue that we may be programmed to die. Recall country singer Tim

McGraw's famous song "Live Like We're Dying." How true! We are all in the process of dying during every minute of every day.

One consequence of the upward change in life expectancy relates to how we are now dying. A century ago, people died mostly from diseases like pneumonia, tuberculosis, influenza, diarrhea, and so on. Today, more people are dying of heart disease, cancer, and strokes. Because of the longer life expectancy, many people are giving in to gradual degenerative-type diseases. Medical advances have resulted in a decrease in deaths from infectious diseases but have brought about more deaths from these gradual degenerative diseases—those in which no one can escape. It seems that organisms, including humans, are programmed to age and to die.

Cells that make up our bodies undergo aging and death just as organisms do. Some cellular death, however, is very normal. When a woman is pregnant, her uterus can be as large as two thousand grams (about seventy-one ounces). Following birth, the uterus recedes to about fifty grams (less than two ounces), resulting in a fifteenfold reduction in size. Much of this decrease is due to cell death. The same can be said for mammary glands during and after pregnancy. Another good example of normal cell death is the formation of the hand during fetal development. Initially the hand looks like a baseball catcher's mitt. Cell death in certain regions of the hand leads to fingers. Fingers do not grow out from some stub; rather, they are carved out by cell death. Certain cells are supposed to die owing to a specific cue (cell signal), all of which is absolutely necessary. However, some cells refuse to die on cue. Often, this situation is cancer. And then some cells die too soon without a cue. These situations are called Alzheimer disease, Parkinson disease, and Lou Gehrig disease. When cells die because of a normal cue, the event is called apoptosis. Apoptosis is *programmed* cell death. The cell dies because it is supposed to die.

A basic question often raised is whether the aging process and the death of cells are correlated to the aging process and the death of organisms. In this regard, an interesting series of experiments was carried out years ago by Leonard Hayflick and his colleagues. This group of researchers showed that cells taken from a human embryo and grown in vitro (in glass) could only undergo approximately fifty, plus or minus ten, cell divisions before dying. The number of cell divisions was calculated by determining the number of times the cell population doubled (figure 32-1). It made no difference how meticulous the researchers were in caring for the cells—that is, keeping the nutrients at optimum levels and not allowing waste materials to build up in the culture. Also, it did not matter what type of cells was used, although most of the experiments were carried out with fibroblasts. Fibroblast cells are involved in fiber formation in the body's connective tissue. One key experiment involved allowing the cells to undergo a number of cell divisions and then placing them into liquid nitrogen for many years. Liquid nitrogen is 196 degrees below zero Celsius (−320.8 °F), a temperature that stops all cell

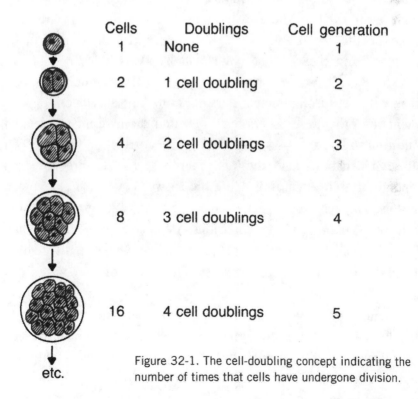

	Cells	Doublings	Cell generation
	1	None	1
	2	1 cell doubling	2
	4	2 cell doublings	3
	8	3 cell doublings	4
	16	4 cell doublings	5
etc.			

Figure 32-1. The cell-doubling concept indicating the number of times that cells have undergone division.

activity. When the cells were eventually removed from the liquid nitrogen and thawed, the resumption of cell division brought the total to about fifty (including before and after freezing). This experimental result was quite surprising. The results suggest that the aging of cells is not so much dependent on time per se, as it is on the number of cell divisions that the cells have undergone. This fifty-cell-divisions phenomenon has become known as the Hayflick limit.

The conclusion resulting from these and other related experiments was that the body's cells contain an inherent aging program. This interesting idea is based upon a substantial amount of data. The hypothesis does not rely on randomness and accidents but on a precise genetically determined program for cellular aging. It was also determined that the program is different for each species.

The Hayflick limit of fifty cell divisions relates to human embryo cells. These cells were taken from embryos that had spontaneously aborted. If these experiments were conducted with cells from people at different ages, would fifty cell divisions still be observed? Initial experiments by Hayflick and his colleagues showed that fewer cell divisions occurred with cells from elderly persons than with cells from younger persons. However, subsequent experiments by other researchers have disputed these results. A related question to be asked is whether there is a correlation between cell division capability and the varying life spans that are observed among different organisms. Here the answer is yes. For example, a tortoise may have a life span of over 175 years, and tortoise embryo cells result in 90 to 125 cell divisions. A chicken, on the other hand, has a life span of 30 years, and chicken embryo cells will divide only 12 to 35 times. It is interesting in itself to find that a chicken can actually live to 30 years if it doesn't become soup first.

Genetic possibilities become apparent when one considers certain human syndromes. Several genetic diseases are characterized by precocious aging. One such disease is called Werner's syndrome. Individuals with this condition begin very accelerated aging at about thirty years old. By age forty or forty-five, affected persons already have the appearance of very elderly people, and not just a little graying. The disorder is due to a recessive allele, and therefore it manifests when the person is homozygous for it. Cockayne syndrome is still another disease that shortens the life span. Dwarfism, mental deficiencies, and a senile appearance characterize this condition.

Even more astonishing is the very rare disease progeria, also known as Hutchinson-Gilford. This mysterious disorder brings about symptoms of aging soon after birth, all of which results in severe aging characteristics by the time the affected persons are only nine to thirteen years old. These victims of precocious aging usually die of "old age" when they are about fourteen years old. Affected persons tend to resemble each other in remarkable ways. Within two years of birth their skin hardens, their hair falls out, and growth is drastically limited. The "children" may stop growing when they reach about thirty to forty-five pounds. Children with progeria become wrinkled and have joint problems early in their lives. Death is often due to the same afflictions that tend to strike elderly people, such as heart disease and strokes. In vitro (in glass) experiments have shown that their cells will only undergo two to ten cell divisions, while normal cell division at their age would be more like thirty to forty. The cause of progeria is not completely clear, but recent investigations indicate the disease to have a genetic basis. The study of progeria may afford researchers new insights into the aging process and maybe an aging clock.

Other experiments have shown that the aging clock, if one truly exists, is located in the nucleus of the cell where the genetic material resides. Cell fusion experiments were designed and conducted to determine the general location of the clock that regulates cellular aging (figure 32-2). Nuclei from cells allowed to undergo ten population doublings (called young) were inserted into cytoplasts (cells without a nucleus) that had been allowed to undergo thirty population doublings (called old). Also, nuclei from cells allowed to undergo thirty population doublings (old) were inserted into cytoplasts derived from cells allowed to undergo ten population dou-

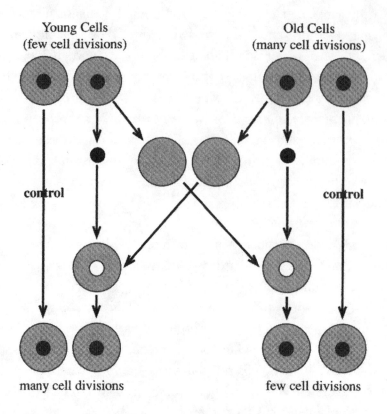

Figure 32-2. A cell fusion experiment designed and conducted with micromanipulation to determine the general location of a "clock" that regulates cellular aging.

blings (young). The results strongly suggested that the nucleus and not the cytoplasm was primarily responsible for cellular aging.

Aging and genetics seem to be correlated when considering many of these experiments and the information from diseases such as progeria and Werner's syndrome. Such evidence brings researchers to the conclusion that the cellular functions of aging and genetics go together. However, in addition to a cellular program for aging, a number of other hypotheses exist for factors involved in the aging process. For example, the accumulation of mutations due to environmental insults occurring over the period of a normal life span might play a role. Such mutations are sometimes called aging hits. Also, our repair mechanisms and immune system seem to weaken as we grow older. And calorie restriction has been shown to slow aging in a number of animals, leading to the conclusion that metabolism is certainly involved, at least in life expectancy if not the life span. All of these factors probably contribute to the inevitable process of aging.

One way to think about this aging process is to place the aging genes into categories that describe the genes as either "private" or "public." Private genes for aging would be those that can differ from individual to individual. These genes would be the ones that make some sixty-year-old people look and act like eighty-year-old people, and other sixty-year-old people look and act like forty-year-old people. These two types of people are obviously genetically different from each other. Public genes would be those that are found in every individual. These genes cause everyone to age and eventually to die regardless of your private genes, your lifestyle, or your luck. Programmed aging never relinquishes its grip on us.

The people of Sardinia, like everyone else, live by the rules of the genetic program for aging, but they show a larger proportion of people who live closer to the life span limits. Sardinia is an Italian island in the Mediterranean Sea where an unusual number of men and women are centenarians (one hundred years old or more). Scientists interested in biogerontology are asking questions about these people. For example, are these remarkable people genetically programmed by their "private genes" toward having a long life? Or is this unusual situation simply due to environmental factors like a life of low stress? Science still has much to learn about getting more people closer to the maximum life span. Sardinia may be a perfect place to study aging. The island is a kind of genetic isolation with few people entering or leaving. It is possible that populations like Sardinia will yield some clues regarding the aging process. Some scientists hope so.

Getting more people to live longer within the limits of our life span is probably a noble scientific objective, but extending the length of the span is quite another matter. The life span (not life expectancy) of a little roundworm named *C. elegans* has been considerably extended by effecting a single gene mutation. Genes have also been modified in yeast cells that extended their life span. Fruit flies, one of our favorite genetic organisms, live an average of 37 days with good care. By modifying a particular gene in experiments, the fruit flies lived an average of 70 days, and some even made it to 110 days. These creative geneticists named the gene in question "Indy" (**I'm not dead yet**). A longevity gene has also been identified in mice that allowed them to have a life span about 50 percent longer than normal. These experiments show that single genes can have a dramatic influence on longevity. Some scientists believe that aging in humans might also be changed. Consequently, research of life-extending genes is an active scientific pursuit. Some scientists wonder whether humans have an "I'm not dead yet" gene or genes. The search is one of those biological Holy Grails. Still other scientists believe that aging and life span cannot be modified. In addition, it may be difficult to retard aging without affecting something else.

Many scientists are probing around in human cells looking for clues to cell division, cancer, healing processes, development, and other cellular activities. One cannot help but think about the possibility of the "Columbus principle" when discussing the aging process. Someone among these investigators could by accident come across the aging clock without actually looking for it. Even more concerning would be this researcher finding a way to change the clock—that is, to set it back. Think about the problems with overpopulation, starvation, retirement, social security, healthcare, insurance, unemployment, poverty, and many other social and political ramifications. Discovering a fountain of youth may not be such a good idea.

FURTHER READING

Comfort, A. 1979. *The Biology of Senescence.* New York: Elsevier North Holland.

Finch, C. E. 2007. *The Biology of Human Longevity.* New York: Elsevier.

Hayflick, L. 1994. *How and Why We Age.* New York: Ballantine Books.

Potten, C., and J. Wilson. 2004. *Apoptosis: The Life and Death of Cells.* New York: Cambridge University Press.

Skloot, R. 2010. *The Immortal Life of Henrietta Lacks.* New York: Crown.

Spence, A. P. 1989. *Biology of Human Aging.* Englewood Cliffs, NJ: Prentice-Hall.

Wolpert, L. 2009. *How We Live and How We Die: The Secret Life of Cells.* New York: W. W. Norton.

REFERENCES

Bohr, V. A., N. S. Pinto, S. G. Nyaga, G. Dianov, K. Kraemer, M. M. Seidman, and R. M. Brush Jr. 2001. "DNA Repair and Mutagenesis in Werner Syndrome." *Environmental and Molecular Mutagenesis* 38: 227–34.

Chu, G., and L. Mayne. 1996. "Xeroderma Pigmentosum, Cockayne Syndrome and Trichothiodystrophy: Do the Genes Explain the Diseases?" *Trends in Genetics* 12: 187–92.

DeBusk, F. L. 1972. "The Hutchinson-Gilford Progeria Syndrome." *Journal of Pediatrics* 80: 697–724.

Epstein, C. J., G. M. Martin, A. L. Schultz, and A. G. Motulsky. 1966. "Werner's Syndrome: A Review of Its Symptomatology, Natural History, Pathologic Features, Genetics and Relationship to the Aging Process." *Medicine* 45: 177–221.

Friedberg, E. C. 1996. "Cockayne Syndrome—a Primary Defect in DNA Repair, Transcription, Both or Neither?" *BioEssays* 18: 731–38.

Goldstein, S. 1971. "The Biology of Aging." *New England Journal of Medicine* 285: 1120–29.

Hayflick, L. 1965. "The Limited *In Vitro* Lifetime of Human Diploid Cell Strains." *Experimental Cell Research* 37: 614–36.

————. 1975. "Cell Biology of Aging." *BioScience* 25: 629–37.

————. 1980. "The Cell Biology of Human Aging." *Scientific American* 242: 58–65.

Hayflick, L., and R. S. Moorhead. 1961. "The Serial Cultivation of Human Diploid Cell Strains." *Experimental Cell Research* 25: 585–621.

Kipling, D., T. Davis, E. L. Ostler, and R. G. A. Faragher. 2004. "What Can Progeroid Syndromes Tell Us about Human Aging?" *Science* 305: 1426–31.

Pollex, R. L., and R. A. Hegele. 2004. "Hutchinson-Gilford Progeria Syndrome." *Clinical Genetics* 66: 375–81.

Pryor, W. A. 1970. "Free Radicals in Biological Systems." *Scientific American* 223: 70–83.

Rusting, R. L. 1992. "Why Do We Age?" *Scientific American* 267: 130–41.

Schneider, E. L., and Y. Mitsui. 1976. "The Relationship between *In Vitro* Cellular Aging and *In Vivo* Human Age." *Proceedings of the National Academy of Sciences* 73: 3584–88.

Wright, W. E., and L. Hayflick. 1975. "Nuclear Control of Cellular Aging Demonstrated by Hybridization of Anucleate and Whole Cultured Normal Human Fibroblasts." *Experimental Cell Research* 96: 113–21.

CHAPTER 33

THE DISCORD BETWEEN EVOLUTION AND CREATIONISM

At the age of twenty-two, after losing his enthusiasm for studying both medicine and religion, Charles Darwin boarded the HMS *Beagle* for a voyage that would continue from 1831 to 1836. The mission of the voyage was exploration, and Darwin was the ship's naturalist. Wherever the ship touched land, he collected samples of exotic plants and animals. He puzzled over them fruitlessly until, in 1835, the *Beagle* reached the Galapagos Islands six hundred miles west of Ecuador. There he heard the vice-governor of one of the islands say that if all the animals from all of the islands were lined up, he could group them according to their islands of origin. This assertion started Darwin thinking. Why should the several Galapagos Islands, which shared the same volcanic geological formations, soil, climate, and topography, have different animals? His efforts to answer this question eventually led to the publication *Origin of Species.*

The animals that intrigued the vice-governor and Darwin were varied. One island had twenty-six species of land birds but shared only one, a bobolink, with the other islands. The other twenty-five resembled South American species. Another island had sixteen species of snails all its own. As a group, the islands had thirteen species of finches, each with its own beak size and shape, lifestyle, and preferred environment. There were ground finches, seed eaters, insect eaters, and fruit eaters. It seemed clear that the beak types suited the various feeding habits, and Darwin began to think about the possibility that one original species had become modified to form the various modern species, each pursuing different feeding strategies for survival. His thoughts grew more definitive as he reviewed the specimens he had collected earlier in the voyage, and as he found similar patterns later on.

Once back in England, Darwin began to build the case for the mechanism underlying evolution. He studied fossils, noting that most differed from living species. He saw how plants introduced into a new area could spread quickly through the countryside. He studied how plant and animal breeders produced new breeds by artificial selection for favored traits, culling those that lacked such traits and breeding those that had them. He argued that nature, too, must select for the traits that aid in survival. For more than twenty years he diligently collected and assembled information on how new species might arise.

Although Darwin did not publish his work during those two decades, he did share it with his scientist friends. This was fortunate because in 1858 another naturalist, Alfred Russel Wallace, sent Darwin a copy of his paper explaining the origin of species by natural selection. Darwin's friends substantiated his priority in the field and convinced him to prepare a joint paper with Wallace for presentation at a scientific symposium later that year. At the same time, they also urged him to proceed with his book, which was published in 1859. Darwin and Wallace had reached the same conclusions about natural selection and evolution quite independently, but the book, and Darwin's immense collection of supporting evidence, brought Darwin most of the recognition.

Evolution has become one of the unifying concepts of all biology.

Not all scientists agree on the precise mechanisms of evolution, but their disagreements should not be taken to mean that they disagree on the process of evolution happening. Evolution is change, and change happens over many generations. The debate is over mechanisms underlying evolution, and debate is one important crux of the scientific method. Because so many observations by scientists of many different disciplines have shown that evolution has taken place, scientists and nonscientists alike sometimes argue about whether evolution should be called a theory or a fact. All science is indeed tentative, but some scientific tenets are very difficult to overturn.

In the public view, the idea of evolution has been especially controversial since Darwin's time. The most reluctant people to accept the theory have been those who take the Bible as their source of truth, sometimes so literally that they believe Bishop Ussher's calculations from biblical writings that the earth was created at 9 AM on October 23, 4004 BCE. For many years in the United States, these people were able to influence state legislatures, school boards, libraries, textbook publishers, and teachers to bar any discussion of evolution from both classrooms and textbooks. Their influence weakened after the famous Scopes trial of 1925 in Dayton, Tennessee. When Tennessee passed a law prohibiting the teaching of human evolution in public schools, John T. Scopes admitted that he taught the "forbidden subject" to his high school biology students. He was found guilty in court and fined, but the case was such an embarrassing spectacle that few states have wanted to ban the teaching of evolution in public schools since then.

After 1925, the opposition to evolution was sporadic until the creationism revival of the 1960s, which was accompanied by a Creation Research Society (formed in 1963), a Creation-Science Research Center (1970), and an Institute of Creation Research (1972). Creationists claim to count some scientists among their numbers, but most of these people are physical scientists, not biologists.

Most creationists no longer seek to ban the teaching of evolution. Instead, they urge school boards and state legislatures to

require that teachers and textbooks give equal attention to both creationism and evolution. They try to strengthen their case by calling
their views "creation science" or "scientific creationism," and more
recently, "intelligent design" (still creationism). They claim that evolution is just as much religion as creationism, and that creationism is
just as much science as evolution. The Creation Research Society
even developed its own biology textbook.

One of the first successes of creationists came when they convinced the California Board of Education to require textbooks to
include creationism as an alternative to evolution. They also gained
influence when Arkansas passed Act 590, a law requiring balanced
treatment for creationism and evolution. However, in 1981 the
American Civil Liberties Union (ACLU) filed suit to invalidate Act
590, contending that it violated the separation of church and state,
and academic freedom. It was argued that it was incredibly vague,
and compliance with it was not possible. The presiding judge ruled
in January 1982 that creationism is a religion, not a science, and that
Act 590 was therefore unconstitutional. The state failed to show otherwise, and evolutionists carried the day. Some observers believe
that creationists greatly underestimated the strength of the scientists' convictions on this matter.

Generally speaking, the public's understanding of the term
theory does not parallel the scientists' use of the term. This contributes to some of the confusion between the thinking of scientists
and nonscientists. A scientific concept is elevated to be a theory only
after voluminous data that support the concept are already amassed
and many contributing facts are available. A theory is not simply a
guess or even a hypothesis.

Evolution is a scientific theory, and creationism is not. Evolution
is not religion, and creationism is religion. In addition, science is a
lively activity characterized by researchers doing experiments that test
hypotheses. Science is not just a body of facts. Creationism, on the
other hand, has no problem-solving or research aspects at all. There is
no part of creationism that is testable because it is entirely a matter of

miracles and faith. The debate is meaningless, and even some creationists seem to know it. So they spend most of their time trying to discredit evolution. In fact, they become exuberant over the controversies among evolutionists even though the controversies concern only the mechanisms of evolution, not whether evolution is occurring. Science is not rigid, and the theory of evolution continues to develop. The competing theories all fall within a certain framework very much agreed upon by all evolutionists and most other scientists.

Creationists sometimes take things out of context. One of the so-called flaws of evolution that gives delight to creationists is the fossil record. They say the record holds no transitional forms, even though many fossils do show the transitions between differing organisms. *Archeopteryx*, a birdlike reptile, is a link between reptiles and birds and an excellent example of a transitional form. And there have been others. Furthermore, they say the record has too many gaps, even though the gaps have been well explained. First of all, very few organisms have ever died under conditions that favored fossilization; much fossil-bearing sediment has been destroyed by erosion; many invertebrate organisms were not capable of fossilization; and paleontologists have yet to find some fossils that may lie in the sediment. Still, if the gaps are real, and not just artifacts of history, they may be due to the actual mode of the evolutionary process—that is, the concept of punctualism. Evolution may occur in large jumps; hence, real gaps may indeed exist. Finally, there is so much evidence for evolution from various disciplines that evolutionists think a good fossil record is hardly necessary. Gaps in the record cannot disprove evolution, despite the opinion of creationists.

People of many religions find no contradiction in evolution because they do not take the scriptures literally. These people can reconcile their religion and evolutionary theory by believing that a Divine Creator was responsible for plants, animals, and humans by means of the evolutionary process. In their system of belief, evolution does not disprove the existence of a Creator.

The evidence for evolution is now overwhelming. Such data

Figure 33-1. Creationism taught in public schools.
© Journal Sentinel, Inc. Reproduced with permission.

have come from biogeography, taxonomy, embryology, physiology, geology, paleontology, genetics, and molecular biology. The origin of life is more difficult to explain, but several hypotheses have been developed, and some data show the feasibility of life beginning on the primitive earth.

Yet the discord will not go away. Millions of creationists live in America. While most scientists respect the beliefs of these people, they also insist that creationism should not be taught in public schools under the guise of science. Since Darwin's time, there have been religious and social controversies over evolution. Today, debates and lawsuits are both part of a climate of opposition between creationists and evolutionists over what should be taught in the public schools concerning our origins.

FURTHER READING

Appleman, P. 1979. *Darwin.* New York: W. W. Norton.
Coyne, J. A. 2009. *Why Evolution Is True.* New York: Viking Penguin.
Darrow, C. 1932. *The Story of My Life.* New York: Charles Scribner's Sons.
Darwin, F., ed. 2000. *The Autobiography of Charles Darwin.* Amherst, NY: Prometheus Books.
de Camp, L. S. 1968. *The Great Monkey Trial.* Garden City, NY: Doubleday.
Ginger, R. 1960. *Six Days or Forever?* New York: Signet Books.
Newell, N. D. 1982. *Creation and Evolution: Myth or Reality?* New York: Columbia University Press.
Quammen, D. 2008. *Charles Darwin on the Origin of Species.* Illustrated edition. New York: Sterling.

REFERENCES

Gurin, J. 1981. "The Creationist Revival." *Sciences* 21: 16–19.
Kettlewell, H. B. D. 1955. "Selection Experiments on Industrial Melanism in Lepidoptera." *Heredity* 9: 323–42.
———. 1956. "Further Selection Experiments on Industrial Melanin in the Lepidoptera." *Heredity* 10: 287–301.
———. 1961. "The Phenomenon of Industrial Melanism in the Lepidoptera." *Annual Review of Entomology* 6: 245–62.
Yunis, J. J., and O. Prakash. 1982. "The Origin of Man: A Chromosomal Pictorial Legacy." *Science* 215: 1525–30.

CHAPTER 34

MOVIES AND
GENETIC IMPLICATIONS

Movies can have many different effects on their viewers. Most movies simply entertain audiences by creating emotional reactions, such as laughter, sadness, enlightenment, rage, weeping, and so on. A few movies additionally have a genetic bent to them. Sometimes the audience is well aware of the genetic side to the plot, and in other cases the audience may not even realize that there are genetic implications. A few of these movies are surprisingly accurate from a genetic standpoint. Other such movies, however, are extremely far from any realistic science.

Several films have already been mentioned in previous chapters, and it is worthwhile to revisit them with a little more explanation. Let's begin with *Twins*, starring Danny DeVito and Arnold Schwarzenegger. The initial inclination before viewing the film was that these two opposites must have been fraternal twins. Even that supposition would be a genetic stretch, but at least possible. Recall that fraternal twins are just like any other siblings, such as brother/brother, sister/sister, or even brother/sister. The same genetics is involved as in other siblings, except that fraternal twins are usually born within a few minutes of each other. Because of the way pairs of genes and chromosomes segregate during the production of sex cells, calculations can show that fraternal twins on the average will have 50 percent of their genes in common with each other. However, the range is technically 0 to 100 percent of the genes being in common. The extremes are probably too remote to have ever happened. Sometimes, however, DNA tests have to be conducted to determine whether newborn twins are fraternal or identical. So what was the story in the movie *Twins*? Danny and Arnold were identical twins, but a bizarre experiment left one with the "good" genes and the other with the "bad" genes. Bizarre indeed, but this plot is, of course, only an entertaining comedy.

Another film relating to comparable genetic concepts discussed earlier, only in the other direction, is *Dave*. Someone in the general public looked exactly like the president of the United States. When the president suffered a life-threatening, debilitating health condition, the presidential advisors brought in "Dave" to fill in for the dying president for political reasons. This cute story is astronomically remote when one seriously considers the genetics. Dave and the president were not even related. The remoteness is again due to the segregation of gene pairs when manufacturing sex cells. Of course, many look-alikes exist in the population, but not that much alike.

Next consider the gripping film *The Elephant Man*. This film is a true story about the plight of a man in London, John Merrick, with the genetic syndrome called neurofibromatosis. The syndrome is caused by a dominant mutation, and the symptoms are quite variable, ranging from mild to extremely severe. Some affected persons may only show a few brown spots on their body, while others will have large tumors on their face and body. John Merrick had a severe disfigurement and spent much of his lifetime as a carnival sideshow attraction. He was badly mistreated, and the theme of the film embodies deep human interest. As previously mentioned, many circus and carnival sideshow people were afflicted with various genetic syndromes, other than neurofibromatosis.

Another interesting film is *The Boys from Brazil.* One might wonder what a title like that has to do with genetics. The story deals with a plan in which ninety-four cloned individuals were grown from Adolf Hitler's cells, taken before his death. The scientific aspects of the film relative to mammalian cloning procedures were amazingly accurate, considering that the film was produced and released long before the breakthroughs in mammalian cloning had actually occurred. In recent years, many different mammals have been cloned, and some scientists believe that humans could indeed be next if not for a moratorium, government regulations, and a widespread attitude that we should not do it. In the film, a great amount of weight seems to be given to the effects of genetics, while almost neglecting possible environmental effects in how these cloned children would react and think. Of course, the film is fictitious but definitely thought provoking.

The Bad Seed, a very old film, prompts people to think about whether criminality has a hereditary component. To briefly review, "Rhoda" was a grade school child with a propensity for killing people. First she did in an elderly lady she knew by pushing her down the stairs. Then Rhoda drowned a grade school classmate at a school picnic because he received a school award that she badly wanted. Finally, a custodian in Rhoda's apartment burned to death while napping on a pile of packing material in the apartment base-ment. Rhoda lit it. In the plot, it was revealed that her grandfather was a murderer, and the "bad seed" supposedly skipped a generation and landed in the genome of sweet little Rhoda. Again fictional, but the film served as a catalyst for the controversial issue of whether there are genes for criminality. In reality, a gene for criminality has never been found. However, some researchers insist that evidence for the inheritance of behavioral traits does exist, some of which could result in antisocial behavior.

Some movie buffs enjoy the films concerning weird mutations, like *Godzilla.* This animal was monstrous, nearly as large as a New York City building. According to the film's plot, the behemoth devel-oped from lizard eggs that were near islands used for atomic bomb testing. Astronomical changes (mutations) occurred causing the development of huge Godzilla in a mere thirty years. Sure! Another example of such rapid organism change can be seen in the movie *Them!* In this film, giant ants also developed in a short period of time. Sure! Mutations do occur and changes in organisms do happen as the result of mutations, but the changes are usually subtle. Many of these small mutations can accumulate over eons of time, eventually creating significant differences. Rapid events leading to an overnight formation of monsters do not happen in nature—only in the movies.

It is amazing how many films produced and released have a genetic twist. *Trading Places* is still another. Can a poor, devious pan-handler be changed into a sophisticated, sharp financier? Two very rich financiers placed a one-dollar bet on whether they could turn this panhandler into an astute colleague. And conversely, whether they could change their reliable financial colleague into a hopeless individual. For a time, they succeeded in creating a major metamorphosis of both persons. The film's story perfectly epitomizes the nature versus nurture controversy. We still do not know much about the extent of genetics versus the environment regarding many complex traits like behavior. Beyond the laughs, the film should make a person think more than a little.

Although filled with interesting science and some of it actually being close to plausible, *Jurassic Park* is still only a novel and a movie of fiction. An island off the coast of Costa Rica served as biological preserve, except the animals roaming it were a group of different prehistoric dinosaurs. The way these extinct animals came about was the result of some ingenious scientific ideas. Bloodsucking insects living during the time of these prehistoric animals were occasionally embedded and preserved in the sap of trees, called amber. Since the insects contained small amounts of the blood of dinosaurs, their DNA could be extracted from this blood and used as the dinosaur's original genetic blueprint. The DNA, in turn, was placed into other animal eggs for incubation and subsequent development. In addition, the geneticists engineered the DNA in such a way that made them all females; hence, the animals could not breed in nature in case they escaped. *Jurassic Park* is a great genetic story up to this point. However, it is the development of the dinosaurs in the eggs of modern-day animals that creates a scientific stumbling block. Even if the dinosaur DNA was intact, which is highly unlikely, no regard was given to the importance of the cytoplasm of the egg—that is, the substance outside of the nucleus of the cell. Before fertilization, eggs are loaded up with RNA messages, metabolic signals, and other molecular machinery in order to begin the development of an organism of the species that provided the egg, in case fertilization of the egg does occur. Be reminded that cloning of animals thus far always requires the nucleus (DNA) to be inserted into an enucleated egg of that species, or at least a very close relative. Nonetheless, *Jurassic Park* is a fascinating movie from a genetic standpoint.

The setting of *Gattaca* is at a time (in the future) when a drop of blood could be taken from the heel of a child immediately after birth and then quickly analyzed for genetic traits. A computer could then relate the probabilities of various health conditions in the person's upcoming lifetime. For example, different analyses of someone's blood might include probabilities as follows: neurological problems 60 percent, manic depression 42 percent, attention deficit disorder 87 percent, heart condition 99 percent, predicted life expectancy thirty years. With the completion of the Human Genome Project and ever-frequent DNA breakthroughs, such a scenario may not be all that far off in the future.

The debate over evolution versus creationism, and more recently intelligent design, never goes away. An excellent film portraying the debate between the two sides, evolution versus creationism, is *Inherit the Wind*. The setting of this film is 1925 in Dayton, Tennessee. The film is a reenactment of the Scopes trial. Tennessee had a state law that prohibited the teaching of evolution in public schools. John Thomas Scopes deliberately taught evolution in the classroom to openly test the law. Various developments led to a subsequent trial in which William Jennings Bryan, a fundamentalist and well-known politician, did courtroom battle with the famed Clarence Darrow, a liberal agnostic criminal lawyer. During the trial, Dayton became something like a revival meeting with a carnival atmosphere. The gist of the defense centered on the issue of the freedom of science. Of course, the jury declared John Scopes guilty. The fine was $100, but was not paid because of an appeal. Soon after, Tennessee dropped the antievolution law. Since then, other such evolution/creationism situations have arisen in other states. In each case, the courts have ruled to protect the separation of church and state.

Genetics can be observed in avenues of entertainment other than the movies. For example, the popular *Seinfeld* comedy show occasionally dipped into the realm of genetics. In one episode, two dwarfs were shown to have a daughter of normal height. This situation is definitely possible if the two parents are dwarfs due to mutations occurring in different genes. For the same reason, two albinos have had, on rare occasions, a normally pigmented child. Such precedents do exist. In another episode, a woman was shown to be bald. Again, women can be bald, albeit at a much lesser frequency than for men. The genetics of baldness in humans is still not definitively worked out. And then there was "bubble boy." These children have a severe immunological disease whereby they do not have the capacity to fight infections like most people. None of these examples in *Seinfeld* were actually fictitious, only the surrounding circumstances.

Genetic concepts are indeed displayed everywhere, even in movies and sitcoms. An astute person will, of course, recognize these genetic twists. An even more astute person will be able to distinguish among reality, possibility, and definite fiction.

FURTHER READING

Crichton, M. 1990. *Jurassic Park*. New York: Ballantine Books.
Lawrence, J., and R. E. Lee. 1955. *Inherit the Wind*. New York: Random House.
Levin, I. 1976. *The Boys from Brazil*. New York: Random House.
March, W. 2005 edition. *The Bad Seed*. New York: Harper Perennial.

CHAPTER 35

GENETIC MISCONCEPTIONS

Misconceptions concerning genetics have always been widespread. Long ago, it was believed that a miniature person was already formed in the head of a sperm. Following the fertilization of an egg, additional growth of the being was all that was necessary to develop into a bouncing baby boy or girl. Even at the time of Mendel's publication in 1866, heredity had a lot of different explanations, many of which were mystical, and some bordered on voodoo. But very important, Mendel showed that genes (called factors) are indeed particles. Genes take up space and have weight just like any other matter. You can keep them in test tubes, store them in a freezer, put them into an envelope, or place them in your shirt pocket. Genes are particles. Once Mendel's laws of inheritance were established and other genetic information became apparent, genes were thought to be like beads on a string. Not quite, however, since we now know that genes are really regions of a very long DNA molecule and not always in juxtaposition to each other; rather, they are scattered throughout our repertoire of DNA molecules (chromosomes).

Not all genes are found in the nucleus of the cell like commonly believed. Most of them are, but a few genes can be found in the mitochondria of the cell. Mitochondria are involved in respiration and the synthesis of molecules called adenosine triphosphate (ATP) that are used in energy relationships. It is now known that thirty-seven different genes are located in the mitochondria of human cells.

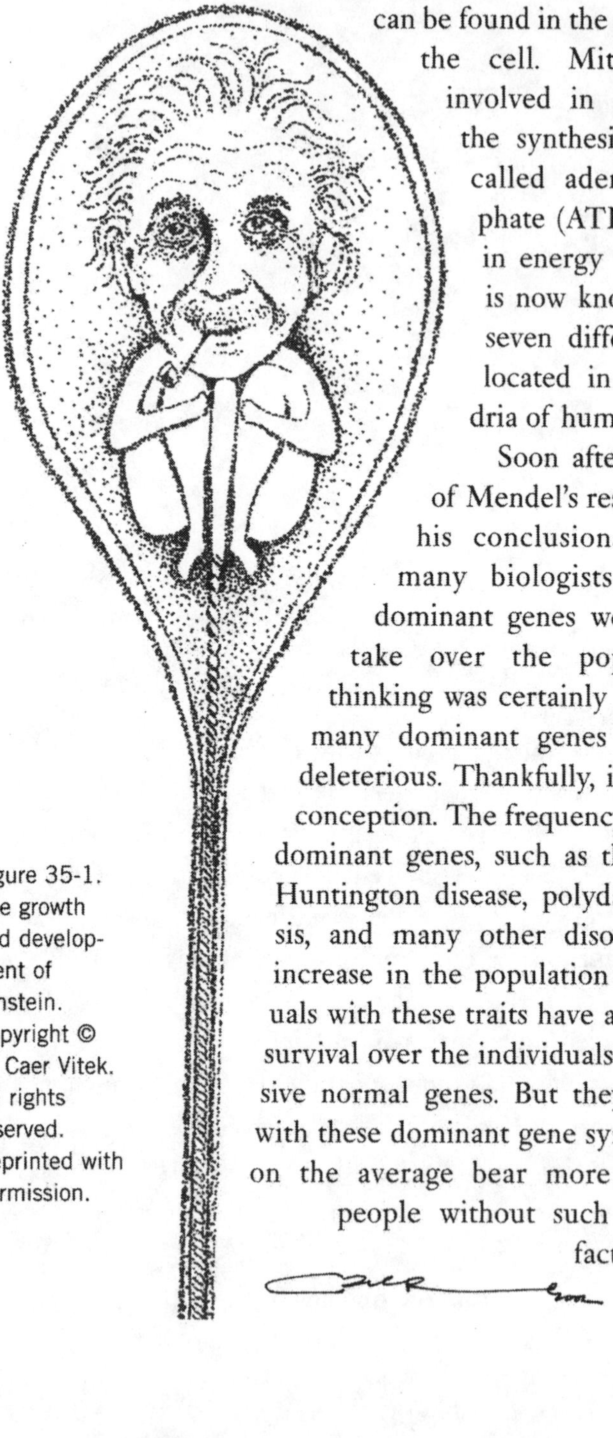

Soon after the discovery of Mendel's research outlining his conclusions on heredity, many biologists thought that dominant genes would eventually take over the population. This thinking was certainly bad news since many dominant genes are extremely deleterious. Thankfully, it was all a misconception. The frequency of deleterious dominant genes, such as those that cause Huntington disease, polydactyly, ichthyosis, and many other disorders, will not increase in the population unless individuals with these traits have an advantage for survival over the individuals with the recessive normal genes. But they don't. People with these dominant gene syndromes do not on the average bear more children than people without such syndromes; in fact, their average number of

Figure 35-1. The growth and development of Einstein. Copyright © by Caer Vitek. All rights reserved. Reprinted with permission.

children is less than normal. Recall that Hardy and Weinberg mathematically showed that the frequency of gene forms would not increase in the population unless they have a selective advantage. Again, deleterious dominant genes do not have selective advantages.

Another misconception of the genetic term "dominant" is that a dominant gene is actually better. But the genetic sense of the term dominant is quite different. A dominant gene is simply the gene (allele) of a pair of genes (alleles) that expresses a characteristic when in a heterozygous condition. The other allele of the pair then is referred to as being recessive. So in a genetic sense, dominant does not necessarily mean better, or that such a gene confers an advantage for the organism. Dying at age forty-five with a neurological disease (Huntington) is not better than normal. Having more fingers and/or toes (polydactyly) is not better than normal. Having a scaly fishlike skin (ichthyosis) is not better than normal. On the other hand, many so-called normal genes are dominant to the deleterious form of the genes, and this situation is indeed a good thing.

One last misconception associated with dominant genes needs to be discussed. On occasion, someone with normal genetic characteristics has a parent with a characteristic or syndrome owing to a dominant gene. The progeny of these parents may still worry about getting the syndrome at a later time, or even passing the gene on to their own children. Generally, this event is not going to happen in either case. Since the gene is dominant and the son or daughter of the affected parent doesn't have the syndrome, he or she must not have the gene. In a very few cases, this explanation doesn't always hold up. Some dominant genetic syndromes have a late onset, and the individual wouldn't know the true situation until later, such as with Huntington disease. Also on rare occasions, a genetic event called incomplete penetrance might occur. This means that the dominant gene is not fully expressing in the individual even though the parent passed the gene on to the individual. In such a case, the individual could, in turn, pass the gene to the next generation. However, these situations are rare, and most of the time if you do not have the dominant characteristic, you are not carrying the dominant gene.

In humans, the ratio of boy babies to girl babies being born is not one to one (50:50). The actual ratio in the United States, and many other countries, is more like 106 or 108 males to every 100 females being born. This may not seem like much of a difference; however, when you consider that the difference is nationwide or even world-wide, the deviation is statistically significant. Understanding the mode of X and Y chromosome segregation and the fertilization event would obviously indicate an expected ratio of one to one, male to female births, but it doesn't quite happen. Scientist think the deviation from one to one may occur because sperm carrying the Y chromosome swim faster and with more motility than sperm containing the X chromosome. Since the X chromosome is so much larger than the Y chromosome, the extra baggage amounts to a 2.8 percent increase. The Y-carrying sperm is more often the winner in the race for the coveted egg.

One often hears that someone got all of his or her genes from his or her father or mother. This heredity is impossible. All of us always

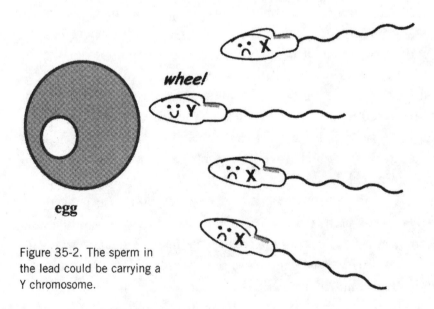

whee!

egg

Figure 35-2. The sperm in the lead could be carrying a Y chromosome.

get half of our genes from our mother and half from our father. It is the way sperm meeting egg actually works. However, the genes from one parent or the other may express better and, therefore, lead to such a misconception. In other cases, people will sometimes refer to their children as F_1s. This genetic terminology is okay. But technically, when they refer to their grandchildren as F_2s, it is not okay. Genetically, F_2 offspring are those produced by crosses between F_1 offspring. Such crosses are permissible in fruit flies and corn plants, but it is a serious societal problem in humans. Such crosses are incestuous. Grandparents probably didn't really mean that their grandchildren are F_2s.

Other misconceptions in human heredity relate to eye color. Over and over again one hears, and even sees in print, that brown eyes are dominant over blue eyes. Not so! Eye color is most likely not a simple Mendelian trait owing to one pair of alleles. The situation is much more complicated since the characteristic is probably due to several pairs of alleles, certainly more than one pair. One important pair of alleles may indeed follow the dominant/recessive mode of inheritance, but other pairs of alleles undoubtedly modify the final eye color expression. I wonder how often a middle school or high school student has been subjected to the erroneous notion that eye color is straightforward dominant/recessive inheritance, looked into the mirror at a pair of fairly brownish eyes, then looked into the more bluish eyes of mom and dad, and finally ran up to the bedroom for a good cry. He or she thought that dad was not really dad. But two people with bluish eyes can have an offspring with brownish eyes, because eye color is not a simple Mendelian trait. Also, two dwarf parents do not always have dwarf children. And two albino parents do not always have albino children. If the two parents are dwarfs because of mutations in different genes, they can have children with normal height. The same situation is true for albino parents due to different gene mutations for albinism. Their children could be normally pigmented.

Some people think that only women pass the extra chromosome 21 to their children, causing Down syndrome. They get this notion from data often reported showing the strong correlation between the woman's age and the occurrence of Down syndrome. Such a correlation does indeed exist, but still the cause of Down syndrome is sometimes due to an extra chromosome 21 being passed by the sperm of the male. Generally though, about 70 to 80 percent of the Down syndrome problem is due to the egg of the female and 20 to 30 percent of the time it is due to the sperm of the male. The notion that only men inherit baldness is also incorrect. The genetic details of baldness are still being worked out, but bald women are also out here, only in a much lower frequency than bald men. And wigs can hide a lot.

Genetic counselors often come into contact with people harboring genetic misconceptions, especially with regard to probability. Many people have a difficult time comprehending the concept of probability. For example, one person seeking genetic counseling authoritatively pointed out that all probability is 50:50. When asked why this should be the case, he replied that everything either happens or it doesn't happen; therefore, it is a 50:50 situation. Another couple, who had a child with a severe recessive Mendelian disease, was told about the well-known 3:1 ratio under these conditions. The parents then left the counselor's office convinced that they already had the "1," and they could now go about having the other "3" without anything to worry about. Parents and potential parents must understand that probability has no memory. In most cases, each occurrence is independent of previous occurrences. People having other notions should stay away from the gambling tables.

Inbreeding among organisms is not *intrinsically* bad. It just looks that way because of the increase in abortions, mentally challenged individuals, and genetic syndromes occurring from inbreeding-type matings. But such occurrences only occur because of the deleterious genes carried by most parents. If they were not carrying such genes, inbreeding would not pose a genetic problem. In a related topic, it can be pointed out that most mentally challenged children are born to couples whereby the two parents are unrelated with normal intelligence. In the first place, most of the people in the population having children have normal intelligence. Second, most mentally challenged individuals do not have children for a number of obvious reasons.

It is easy to see why many people think that certain combinations of genes are passed on to offspring. Dark hair and brown eyes or light hair and blue eyes are a couple of examples. During sex cell formation, most genes segregate independently from each other. Even if two genes are next to each other on the same chromosome, over time they will separate from each other owing to the crossing over of chromosomes—that is, exchange of chromosome parts. However, some populations have a high frequency of certain genes and gene combinations, and continual mating within the population will tend to keep those gene combinations intact. Random mating of individuals among different populations would, over time, reduce the frequencies of these gene combinations.

Some misconceptions relate to the genetics connected to abortion, both spontaneous abortions (miscarriages) and elected abortions. First of all, the frequency of spontaneous abortions is not rare, as some people believe. Clinical data have shown at least 15 percent of all egg plus sperm conceptions abort. But the real figure may be even higher. Extremely early abortions might occur that are not counted or even known to have happened. It is also thought that elected abortions of embryos will reduce specific deleterious genes in the population. Calculations show, however, that the frequency of the culprit gene in the population following elected abortions is lessened very little, if at all. One reason is due to the fact that most of these genes are recessive, and they continue to be harbored in heterozygous (carrier) people. This situation is especially true for very rare genes, which most of them are. Second, normal genes mutating to deleterious genes is always ongoing. And then there is the male parent sometimes blaming the female parent for experiencing a number of successive spontaneous abortions. They ask the question, what is wrong with her? Genes come in pairs, and it takes the contribution of two parents for most syndromes, or sometimes just the contribution of the male.

Some weird impressions circulate with regard to radiation, too. One such notion is that if your gonads are protected from radiation, like the lead shield at the dentist's office, you are completely safe from the effects of radiation. Protection of the gonads is very important, and it surely helps to prevent unwanted mutations in the gonad area. But mutations can occur in any of our cells (called somatic cells), not just in the gonads. Somatic mutations can often be the cause of cancer. Some of the public is against the irradiation of foodstuffs used to kill bacteria, fungi, and other unwanted microorganisms. They think the irradiation will make the food radioactive. Wrong! Irradiated food is no more radioactive because of the treatment than your arm that was x-rayed for a possible fracture.

Finally, let's discuss misconceptions concerning natural selection and evolution. Mutations are not all bad. They are the source of variation. Many of the great hereditary attributes that you enjoy are due to good mutations. In order to take advantage of the few good mutations, we have to endure the occurrence of many bad mutations in the population. Also, organisms cannot change their heredity because of the environment that they encounter. We are stuck with what we have. However, populations can change gene frequencies over time; individual organisms in the population cannot. Darwin did not invent evolution. This concept was proposed, discussed, and debated well before Darwin's time. Credit goes to Darwin for his ideas of natural selection, deemed to be the mechanism by which evolution occurs. And humans did not evolve from monkeys. Evolutionists believe that the two species evolved separately from a common ancestor, a weird-looking animal that appears nothing like a monkey or a human.

REFERENCES

Hardy, G. E. 1908. "Mendelian Proportions in a Mixed Population." *Science* 28: 49–50.

Mendel, G. 1866. "Experiments in Plant Hybridization." English translation reprinted in *Classical Papers in Genetics*, edited by J. A. Peters, 1–20. Englewood Cliffs, NJ: Prentice-Hall, 1955.

Weinberg, W. 1908. Translated in English in S. H. Boyer, *Papers on Human Genetics*. Englewood Cliffs, NJ: Prentice-Hall, 1963.

CHAPTER 36

GENETICS AND POLITICS CROSSING PATHS

Genetics makes a tremendous impact on medicine, agriculture, psychology, industry, and even philosophy. Anyone who needs further evidence of the importance of genetics has only to browse through the pages of popular magazines, such as *Time* and *Newsweek*, among others, and daily newspapers. They frequently describe new developments and applications of genetics.

It should surprise no one that genetics also affects politics, and that politics affects genetics. Genetics exerts a great amount of influence on how people should live, the choices they must make, and how they define themselves and each other. These are all issues traditionally in the domain of politics, philosophy, ideology, and religion that arouse strong feelings. Issues relating to genetics can affect the outcomes of elections, and the outcomes of elections can dictate what kind of genetic research will receive public support and funding.

The number of politically sensitive genetic issues increases yearly. Programs such as genetic screening, genetic counseling, worker protection regulations, stem cell research, genetic engineering, genetically modified foods, among others, have stirred up controversy. Politicians have debated aspects of reproductive biology that involve abortion, birth control, in vitro fertilization, artificial insemination, embryo transfer, surrogate mothers, and cloning. Educated and informed people disagree over biological determinism, measurements of intelligence, many aspects of the nature-nurture controversy, and Darwinian evolution.

Several United States government agencies are directly concerned with science. These agencies seek legislation and funds to support their own activities and to support scientific research throughout the country. They include the National Science Foundation (NSF), the National Institutes of Health (NIH), the United States Department of Agriculture (USDA), the Environmental Protection Agency (EPA), the Department of Energy (DOE), the Food and Drug Administration (FDA), and others. Many American scientists often feel that they are pressured to take political stances that agree with current government positions, simply because they depend on political agencies for their funding. Even scientists employed on university campuses and by industry depend at least indirectly on government funding, so it is difficult to be apolitical. Some scientists and science organizations lobby in an effort to persuade members of Congress and agency officials to support unpopular or novel paths of research.

Researchers also influence the politics of science in other ways. Each year, between two thousand and three thousand prestigious scientists are called upon by Congress and government agencies to give advice and various other forms of testimony on technical issues. As a result, scientists can have an effect on legislation. Scientific input, often relating to genetics, has influenced whether laws pass or fail. Scientific input can also shape military plans (as it did in the case of the atomic bomb) and administrative policies in health, education, and civil rights.

Cases worldwide can be pointed out in which a lack of humanizing influences on the science-politics combination has led to disaster. More often, the disaster occurred because ideologues collaborated with one or two scientists whose ideas matched the ideology too closely, and then closed their minds to other points of view. This is what happened in Nazi Germany where National Socialist ideologues allowed hereditarian notions to take a central place in German policy. With all other information shut out, these views quickly led to drastic eugenic measures and the eventual death of millions of people. Since the Holocaust, scientists have been more vocal and less inclined to simply stay in their laboratories when they believe their governments are making mistakes.

Some countries have considered their scientists to be dangerous people because science does not necessarily respect their ideology. Yet other persecutions of scientists are sometimes based on their science, or their political views or activities, especially if they support political opponents or systems they consider more favorable to the values of science. Consider the famous geneticist H. J. Muller. In his time, many did think of him as being dangerous, for he was a very controversial personality all of his life. Muller served as faculty advisor to the Communist students at the University of Texas; he advocated sperm banks for only bright men, along with other eugenic practices; and he was an active atheist. However, he openly opposed the Lysenko movement in the Soviet Union. The Lysenko movement in the Soviet Union is another infamous example of how dangerous a science-politics mesh can become, when the danger is that of error and not truth. There is no better illustration of how a program based on misunderstood genetics and the meddling of politics could cause disaster. Lysenko nearly destroyed Soviet agriculture in the years following World War II.

The incredible story of Lysenkoism lasted almost twenty years. This erroneous concept began in 1948 when Trofim D. Lysenko became director of the Soviet Academy's Institute of Genetics and president of the Lenin Agricultural Academy with the help of theoretician I. I. Prezent and Stalin's absolute dictatorial power. Lysenko believed in Lamarckian evolution; that is, he believed that characteristics acquired by organisms during their lifetime could become heritable and passed to succeeding generations. Furthermore, he insisted on his "physiological theory" despite a long history of genetic experiments showing that his ideas were false. Scientists now know of a phenomenon called epigenetics, whereby DNA can, to some extent, be conditioned to express differently. However, most of Lysenko's supporting data were very weak, and some it seems to have even been faked to appear more impressive. How could Lysenko achieve his position of power? His extreme environmentalist views seemed more relevant to Marxism and Communist political doctrines than did the more generally accepted ideas. In fact, Lysenko boldly denounced all classical genetics because it contradicted those doctrines, and his power quickly destroyed classical genetics in the Soviet Union. Textbooks were rewritten and curricula changed because of these erroneous views. Unbiased, pure genetic research practically disappeared.

Lysenko's power was enhanced by his promises to use his expertise in genetics to greatly expand the productivity of Soviet agriculture, just when the need for increased crops was critical. Stalin believed him, and the political apparatus supported Lysenko's control of biology and agriculture and his suppression of other scientific ideas. Soviet scientists who dared to oppose the Lysenko movement were demoted, arbitrarily dismissed, jailed, exiled, and according to some Western observers, even executed. One famous plant geneticist, N. I. Vavilov, was accused of being a British spy and sent to a Siberian labor camp where he eventually died.

Lysenkoism rejected the sound breeding practices and hybridization techniques being developed in the West. This decision was a drastic mistake that caused the Soviet agricultural productivity to decline drastically. Meanwhile, Western agriculture and genetics achieved great advances. DNA was identified as the genetic material, and its molecular structure was worked out. Molecular biology was born, and the pace of genetic discovery quickened.

The Soviet Union had a strong tradition of biological and agricultural development before Lysenkoism. Subsequently under

Nikolai Krushchev, the situation improved somewhat. Leonid Brezhnev and Aleksey Kosygin finally restored genetic research to its original status of free inquiry. Lysenko was discredited and removed from control, leaving only a new word, "Lysenkoism," to mark his place in history. The Soviet Union's return to sound genetics has been welcomed by all geneticists. The Lysenko episode was a sad affair for science, politics, and human rights. Now Soviet scientists are catching up with the rest of the world, especially in molecular genetics.

Genetics, and science in general, responds to political pressures in places other than the Soviet Union. In the United States, these pressures are usually expressed in legislation and funding, as after the 1957 launchings of the Soviet satellites, *Sputnik I* and *Sputnik II*. The United States reacted by putting billions of dollars into science education, scientific instrumentation, space research, and other scientific and technological areas. In 1961, President John F. Kennedy committed the nation to putting a man on the moon within that decade. The result was an explosion of scientific advances in all areas, and man did in fact reach the moon. Another example of how political commitment can support science came from the "war on cancer" prompted by Congress and President Richard Nixon in the early 1970s. The "war" resulted in large amounts of money being placed into cancer and related genetic and cellular research. This research has paid off in a much-improved understanding of cancer, basic biology, and molecular genetics.

These examples illustrate what may be the only proper way for a government to control research. Certainly, governments have the right to decide what types of research should be encouraged through funding. And if the government provides the money, it has a responsibility to monitor the research, at least in the sense of ensuring that public monies are well spent. Agencies exert this control largely by screening research proposals, but also by channeling funds for training students into certain fields and supporting new construction at particular universities.

Public participation can and should also affect American research. For example, a massive study designed to detect children as XYY individuals, which would then allow the monitoring of their behavior over a long period of time, was abandoned because of mounting pressures from both inside and outside of the scientific community. Many of the criticisms centered on the dangers of stigmatizing people without any adequate justification. Research in reproductive biology is always in the public eye, and genetic research on humans, especially within the realms of intelligence and race, genetic engineering, and cloning, has many vocal opponents in the public. The public strongly favors science in general, but many people object to anything that relates to the creation of new life-forms or threatens traditional practices related to religion and one's philosophy of life.

The government seems much more reasonable when it reacts to possible hazards, as in the federal controls on recombinant DNA research. However, it should be noted that it was not the government or the general public, but the scientists involved, who first drew attention to the possible dangers inherent in this kind of research. In an almost unprecedented way, scientists imposed a moratorium on their own work, devised safety guidelines, and called for regulation. This movement led to a great deal of debate, testimony, and lobbying in the mid-1970s. Legislation for liability and strict fines for researchers who ignored the guidelines were seriously proposed. The "Friends of DNA" argued for less control on the research. And some cities, such as Cambridge, Massachusetts, proposed their own controls. The Cambridge city council wanted to forbid scientists at Harvard University from engaging in recombinant DNA work within the city without its approval.

Some opposition to genetic research is based on concerns for safety. There is also a faction of the public that is completely opposed to human genetic research because of religion or their personal brand of ethics. Many of these opponents of research are worried that new discoveries could be twisted to hamper progress toward racial equality. On the other hand, other geneticists argue that abuses arise not from genetic research but rather from the lack of genetic research. Genetic research, not its absence, was what helped refute the Nazi concept of racial superiority. Proponents of the freedom to research such areas further point out that we must acquire more genetic information regarding human potentials if we are ever to achieve equal opportunity for all. Biologically diverse people may not be able to develop to their optimum levels under the same environmental conditions.

Most scientists and politicians agree that politically motivated suppression of scientific research is not a productive route for society. Scientists must guard against the misinterpretation of data, but they should not fail to seek the data in the first place. It is crucial to make everyone aware of research results. Future decisions based on research could greatly affect society. And everyone in our society needs to be politically aware, and more important, to fully understand the issues.

FURTHER READING

Kamin, L. 1974. *The Science and Politics of IQ.* New York: Halsted.

Lappe, M. 1979. *Genetic Politics: The Limits of Biological Control.* New York: Simon & Schuster.

Lipkin, M., Jr., and P. T. Rowley, eds. 1974. *Genetic Responsibility.* New York: Plenum.

Medvedev, Z. A. 1969. *The Rise and Fall of T. D. Lysenko.* Translated from the Russian version by I. M Lerner. New York: Columbia University Press.

Prudhomme, J. O. 2006. *Anticipating Human Genetic Technology.* Parkersburg, WV: Interactivity Foundation.

Soyfer, V. N. 1994. *Lysenko and the Tragedy of Soviet Science.* New Brunswick, NJ: Rutgers University Press.

REFERENCES

Cohen, B. M. 1982. "H. J. Muller: A Memoir." *Journal of Heredity* 73: 477.

Darby, W. J. 1980. "Science, Scientists, and Society: The 1980s." *Federation Proceedings* 39: 2943–48.

Muller, H. J. 1927. "Artificial Transmutation of the Gene." *Science* 66: 84–87.

CHAPTER 37

CONCLUSIONS

The discussions in this book have focused on some of the many ways in which heredity interrelates with societal activity. Many more roles of genetics exist than are found in this book. The discipline of genetics and its related discipline of molecular biology are advancing at a fantastic pace. A great amount of excitement exists in the scientific community that also spills over into much of the public. People should be alerted to the course of events in genetics that lead not only to beneficial advances but also to controversial issues and problematic situations. By delving into these situations, it becomes quite evident to everyone that biological outcomes are not always definitive; that genetic information is not always easy to interpret; and that in some cases we may not yet have enough information to make sound decisions.

Heredity indeed personally affects everyone, and on the levels of the family, the community, the nation, and the world. Certainly this relationship is true when one regards genetic diseases that directly affect the patient and the patient's family. Beyond these unfortunate situations, however, most people throughout the nation pay taxes for research on genetic diseases and the treatments of them. With the decline of infectious diseases and increased longevity, some of our most serious remaining health problems, such as heart disease and cancer, have become widespread. These health problems also have genetic components. Some researchers feel that the genes owned by individuals influence all diseases. In addition, national and international agencies are concerned with family planning, the rights of the handicapped, biomedical ethics, and an array of other serious medical issues, all of which play a role in questions about genetics.

Genetics and society interact in numerous ways. Medical science confronts genetic disorders, cancer, fetal diagnosis, genetic counseling, mutations, and chromosome aberrations. Society is concerned about genetic damage resulting from radiation, drugs, and chemicals in our environment. Society is concerned about reproductive issues such as abortion, inbreeding, artificial insemination, in vitro fertilization, embryo transfers, sex selection, and cloning. Human behavioral studies are becoming increasingly interesting, and many behavioral traits have been assigned some genetic involvement. Schizophrenia, mental depression, mental deficiencies, intelligence, and even alcoholism and criminality are the subjects of genetic research.

In agriculture, straightforward genetic techniques consist of selective breeding and hybridization. Many of the newer techniques being applied to agriculturally important plants and animals constitute genetic engineering, a subject of very wide interest. This interest is especially lively when it comes to recombinant DNA work that places genes from one species into the DNA of a completely different species. Such genetic engineering techniques are also being applied to humans. This research has worried some people, and genetics has found itself sometimes embroiled in controversies with the public. Embryonic stem cell research is another prime example. Very little science fiction is still science fiction.

Genetics has also met religion and politics in the evolution versus creationism debate and in the discipline of sociobiology, which attempts to explain much of behavior and societal structure in terms of heredity. Genetics has met politics in such worldwide tragedies as Hitler's infamous and misguided eugenics against Jews, Gypsies, and Eastern Europeans. The Lysenko incident in the Soviet Union is another example in which too much reliance was placed on ideological beliefs, and this caused a nearly complete collapse of Soviet agriculture and the dependence on North American wheat supplies. The mere mention of wheat brings to mind the wheat breeder Norman Borlaug, who won the Nobel Prize for Peace because of the close relationship between his improved wheat varieties to world food supply and, in turn, world peace.

Controversy surrounds many of the topics involving genetics in society. All individuals do not judge the moral, ethical, and legal aspects of these problems in the same way. Very often, the achievements of technology are far ahead of society's capacity to accommodate them, and scientists do not always know the extent of the social impact that will result from information already available. Almost everyone agrees that some of the problems are too important and their consequences too great to be left to scientists alone. Therefore, everyone needs to be somewhat educated about these matters. Society as a whole should take an interest in them. Many critical decisions regarding the interaction between genetics and society will certainly have to be made in the future, and the direction of these decisions could greatly affect humanity. Such concern may not be an overstatement. Some of these decisions could actually change the course of history.

FURTHER READING

Bodmer, W. F., and L. L. Cavalli-Sforza. 1976. *Genetics, Evolution, and Man.* San Francisco: W. H. Freeman.

Kieffer, G. H. 1979. *Bioethics: A Textbook of Issues.* Reading, MA: Addison-Wesley.

Kowles, R. V. 1985. *Genetics, Society, and Decisions.* Glenview, IL: Scott Foresman.

Mange, A. P., and E. Johansen Mange. 1980. *Genetics: Human Aspects.* Philadelphia: Saunders College.

Oosthuizen, G. C., H. A. Shapiro, and S. A. Strauss, eds. 1980. *Genetics and Society.* New York: Oxford University Press.

Stine, G. J. 1977. *Biosocial Genetics: Human Heredity and Social Issues.* New York: Macmillan.

GLOSSARY

ABH blood system. A major blood system. Its identification is based on the occurrence of agglutination of the red blood cells when bloods from incompatible groups are mixed. The blood types (A, B, AB, and O) are named based on the particular antigens found on the red blood cells and the antibodies in the plasma.

ABO blood group. See **ABH blood system**.

abortion. The expulsion of an embryo or fetus from the uterus before it is sufficiently developed to survive.

achondroplasia. A dominant form of dwarfism in humans.

additive effects. The situation in which a group of genes affects the same phenotypic characteristic. Each gene adds to the effects of the others.

adenovirus. Any group of spherical viruses that infect a number of mammalian species, including humans.

AID. The abbreviation for artificial insemination by a donor. Artificial insemination is the impregnation of the female without direct sexual contact.

AIH. The abbreviation for artificial insemination by the husband.

albinism. In animals, the absence of the pigment melanin in the skin, hair, and eyes.

alkaptonuria. An inherited metabolic disorder that is due to an autosomal recessive gene. Alkaptonurics have excessive amounts of homogentisic acid in their urine because they lack the specific enzyme for breaking down the substance. The colorless homogentisic acid is oxidized to a black pigment when exposed to air.

alleles. Two or more alternative forms of a gene that occupy the same relative sites, or loci, on homologous chromosomes.

Alzheimer disease. A common type of dementia, progressive and degenerative.

Ames test. Also known as the *Salmonella*-mammalian microsome assay; a test to determine the mutagenicity of various chemicals.

amino acid. The basic chemical units of polypeptides, which are organic compounds containing an amino group, a carboxyl group, and a variable side chain.

amniocentesis. A technique in which a sample of amniotic fluid is removed from the amniotic sac of a pregnant woman by needle puncture. This fluid and the suspended cells in it are then used for prenatal detection of fetal disorders.

amyotrophic lateral sclerosis. A degenerative form of motor neuron disease affecting voluntary muscular movement.

androgen. Any substance with male sex hormone activity.

aneuploid. A chromosome number in a cell or organism that is not an exact multiple of the haploid or basic number.

antigen. Any foreign substance or large molecule that can stimulate the production of specific neutralizing antibodies.

antiparallel. Describes two molecules or parts of two molecules that lie parallel to each other but point in opposite directions. The DNA molecule serves as a good example.

apoptosis. A form of cell death that is a normal path of growth and development.

artificial insemination. The impregnation of the female without direct sexual contact.

artificial selection. The type of selection in which humans choose the genotypes of a given species that will contribute to the gene pool of succeeding generations.

asexual reproduction. Any mode of reproduction that does not involve the union of genetic material from two sexes or mating types.

autosome. All of the chromosomes other than those designated as sex chromosomes.

Barr body. The inactive X chromosome generally observed appressed to the nuclear membrane in the nuclei of mammalian somatic cells.

base pair. A pair of nitrogenous bases that laterally join the component strands of the DNA double helix.

benign tumor. An abnormal localized population of proliferating cells in an animal. The tumor does not spread from the original site to another part of the body.

biogerontology. The study of aging with the use of biological techniques.

biological determinism. The controversial concept that humans are very much the products of their genes.

blastocyst. An early stage of mammalian development in which the cell mass is a hollow ball-shaped structure.

Bombay phenotype. A blood phenotype in which a functional substance called H is absent. This substance is the precursor in the synthesis of the A and B blood antigens.

brachydactyly. An abnormality in humans caused by a dominant gene. The phenotype consists of markedly shortened and thickened fingers.

buccal cell. A cell from the mouth cavity or the inner cheek.

Burkitt lymphoma. A monoclonal malignant proliferation of B lymphocytes affecting the jaw and associated facial bones.

carcinogen. A substance capable of inducing cancer in an organism.

carcinoma. A cancer of epithelial tissues (skin cancer).

carrier. An individual who is heterozygous for a normal gene and one of its alleles. The latter is not expressed because of complete dominance of the normal gene.

cell. The basic structural and functional unit of life, and the smallest membrane-bounded unit of protoplasm produced by independent reproduction.

cell cycle. The life cycle of an individual cell; specifically, the timed sequence of events occurring in a cell in the period between mitotic divisons.

cell fusion. The experimental formation of a single hybrid cell by the physical union of two separate body cells.

chloroplast. A chlorophyll-containing organelle in plant cells involved in photosynthesis.

chromatin. The DNA-protein complex of chromosomes in higher forms of life making up the substance of the nucleus.

chromosome. A DNA-protein complex that is the structure for an array of genes. In higher forms of life, it is essentially a molecule of DNA with other associated molecules.

chromosome aberration. Any change in chromosome number or basic chromosome structure.

chromosome theory. The theory that describes chromosomes as the vehicles for genes and genetic information.

cleavage. The process in which a fertilized egg cell divides mitotically to give rise to the initial cells of the developing embryo.

cloning. The technique of developing large numbers of cells or an individual from a single ancestral somatic cell.

Cockayne syndrome. A disease characterized by dwarfism, mental retardation, and senility, among other symptoms. Generally the life span of such individuals is greatly shortened.

codominance. The situation in which alleles produce an independent effect. In a heterozygote, there will be full expression of both alleles.

combining ability. The average performance of a strain in a series of crosses.

concordance. A term used in twin studies to describe the situation in which both members of a twin pair exhibit a certain trait.

consanguinity. Genetic relationship by descent resulting from a common ancestry, at least through the preceding few generations.

controlling elements. See **transposon**.

correlation coefficient. The measure of the intensity of an association. It is an unbiased estimate of the corresponding degree of association between two variables in a population.

cosmic radiation. Variously charged particles that reach the earth from outer space.

creationism. The belief that the world was created by a supernatural power with all of its life-forms intact just as they are now.

Crigler-Najjar syndrome. A stunting of body growth and mental development in humans due to a deficiency of thyroid hormones.

crossover. The reciprocal exchange of corresponding segments of genetic material between homologous chromatids. This results in the recombination of alleles.

cryopreservation. The treatment and storage of eggs, sperm, somatic cells, or embryos by freezing at very cold temperatures to preserve them for later purposes.

cystic fibrosis. A recessive hereditary disease characterized by the production and accumulation of large amounts of viscous mucus in the lungs.

cytoplasm. The protoplasm of the cell outside of the nucleus containing the organelles of the cell.

Darwinian. Synonymous with natural selection and adaptation.

daughter cells. The two cells resulting from division of a single cell.

deficiency. In a chromosome sense, the same as a deletion. The absence of a segment of the genetic material in a chromosome that may range from a single nucleotide to a significant part of a chromosome.

deletion. See **deficiency**.

deoxyribonucleic acid. Abbreviated DNA. A molecule composed of nucleotide units, usually double stranded and helically coiled. The molecule is the principal material for the storage of genetic information in most organisms.

differentiation. A sequence of changes that are involved in the progressive specialization and diversification of cell types.

diploid. The state of having two complete sets of chromosomes whereby the sets are homologous to each other. It is the typical number of chromosomes for a species, generally symbolized as $2n$.

DNA fingerprinting. See **DNA profiling**.

DNA probe. Tagging a DNA molecule in some way to identify and follow it in various types of experiments.

DNA profiling. A biotechnology that detects differences in the number of copies of certain repeated DNA segments among individuals. Used to rule out or establish identity.

Edwards syndrome. A highly lethal disorder resulting from the presence of three doses of chromosome 18.

Ehlers-Danlos syndrome. A family of hereditary diseases characterized by overelasticity of the skin and by excessive extensibility of the joints.

electrophoresis. A technique that allows for the migration of molecules or particles in an electric field and the detection of these molecular migrations.

emasculation. In a botanical context, the removal of the anthers from a flower.

embryo. A young organism in the early stages of development arising from a fertilized egg.

embryo transfer. The removal of an embryo from one organism and the subsequent placement of the embryo into another organism.

endonuclease. An enzyme that can internally cut the polynucleotide strands of DNA.

enucleation. The removal of the nucleus of a cell.

environmentalist. In one sense, an extreme proponent of the concept that experience is the most important ingredient in the expression of certain traits, especially behavioral traits.

enzyme. A protein that acts like a catalyst. It regulates the rate of a specific chemical reaction in the cells of organisms, without being consumed in the reaction.

epigenetic. A layer of information placed on a gene that is a modification other than a change in DNA sequence.

eugenics. Improvement of the phenotype of a person by such means as selective breeding, chemical administration, surgery, or other biological methods.

extrachromosomal inheritance. Describes the hereditary processes that are not due to the standard chromosomes, which are nuclear; rather, this inheritance is usually cytoplasmic.

extracorporeal gestation. The period in which an environment is provided in some artificial way for developing the offspring after conception, such as with a machine.

F_1. The first filial generation—that is, the generation of individuals produced by the first parental generation being considered in any pedigree or mating scheme.

F_2. The second filial generation—that is, the generation of individuals resulting from self-fertilization or intercrosses of the F_1 individuals.

Fallopian tube. The oviduct in the human female that conveys the egg, and following fertilization the young embryo, to the uterus.

fertilization. The union of a male gamete with a female gamete to produce a zygote.

fetus. The prenatal stage, usually in humans, that refers to the time between the embryonic stage and birth.

fibroblast. Spindle-shaped differentiated cells involved in fiber formation in connective tissues.

follicle. A vesicle-like structure in the mammalian ovary within which the ovum develops.

forensic genetics. Genetics in its relation to law.

founder's principle. The establishment of a new population by a very small number of individuals who isolate themselves from the parent population in some manner. These individuals may not be representative of the gene frequencies expressed within the larger parent population.

galactosemia. A recessive genetic disorder in humans resulting from an inability to metabolize galactose.

gamete. A mature reproductive cell capable of fusing with a cell of similar origin but of opposite sex.

gamete intrafallopian transfer. An infertility treatment in which sperm and oocytes are placed in a woman's uterine tube.

gangliosides. A family of chemically complex lipids that is very prevalent in some cells and can cause certain genetic syndromes.

gene. The functional unit of inheritance. The gene is a segment of a DNA molecule that codes for messenger RNA and the synthesis of a single polypeptide or for a functional RNA molecule.

gene flow. The movement of genes within a population or from one breeding population to another through the dispersal of gametes or zygotes.

gene pool. The sum total of all the genetic information possessed by an interbreeding population at a given time.

genetic code. The sequence of three adjacent nucleotides in the DNA and the messenger RNA that specifies the twenty amino acids translated into polypeptide chains.

genetic drift. Change in gene frequencies of a population from one generation to another due to random chance fluctuations or sampling errors; mostly evident when the population is small. Also called random drift.

genetic engineering. The intentional manipulation of the content and/or organization of an organism's genetic material.

genetic load. The number of unexpressed lethal or detrimental genes maintained in a population or gene pool.

genetics. The study of heredity—that is, the study of variation in organisms, which traits are inherited, and how they are inherited.

genitalia. The sex organs directly involved in the fertilization process of sexual reproduction.

genome. A single set of chromosomes or chromosomal genes inherited as a unit from one parent—that is, a complete set of genes. Sometimes "genome" is used to refer to a complete gene complement regardless of the chromosome number.

genomic imprinting. Differing of the phenotype depending upon which parent transmits a particular allele.

genomics. The study of the function and interaction of many genes.

genotype. The genetic constitution of an individual as distinct from the appearance of the individual (phenotype).

germ cell. A cell from the line of cells that gives rise to gametes, which are capable of fertilization in the sexual reproductive process.

Hardy-Weinberg principle. The notion that the frequency of alleles and the resulting genotypes will remain constant within a large randomly mating population from one generation to another, unless certain forces are acting upon the population.

Hayflick limit. The concept that a population of embryonic cells will undergo a specific number of divisions, and therefore doublings, before becoming senescent and dying. In humans, the limit is estimated to be approximately fifty cell doublings.

hemizygous. The condition in which only one allele of a pair is present in the diploid organism, as in the case of X-linked genes in the mammalian male.

hemoglobin. Protein molecule found in the red blood cells and a transporter of oxygen. The molecule is composed of two different pairs of identical polypeptide strands and an iron-containing heme group.

hemophilia. A hereditary disorder that is due to an X-linked recessive gene. The disorder manifests itself as a tendency to bleed profusely even from small wounds because of the lack of normal blood-clotting mechanisms.

hereditarian. A term sometimes used to describe those persons who are extreme proponents of heredity as the underlying basis of traits, especially behavioral traits.

heredity. The overall process of transmitting genes for traits from parents to offspring, usually resulting in a resemblance among individuals related by descent.

heritability. A quantitative measure of the extent to which the total phenotypic variation of a trait is the result of genetic factors.

hermaphrodite. An individual with both male and female reproductive organs. More technically, both ovarian and testicular tissues are present in the one individual.

heteradelphian. A person born with extra limbs due to developmental errors. An example of teratogenesis.

heteroplasmy. Mitochondria in the same cell having different alleles of a particular gene.

heterosis. The state in which being heterozygous for one or more genes is superior to being either of the homozygotes. Heterosis usually results in hybrids that have more vigor and are more fit than the homozygotes.

heterozygous. The state in which an individual carries both members of a pair of alleles at a given locus.

HLA. See **human-leukocyte-associated antigens**.

homeostasis. A steady state maintained in an organism usually due to genetic resiliency.

homogentisic acid. A compound derived from the metabolic breakdown of the amino acid tyrosine. A buildup of this substance occurs in the urine of persons with the hereditary disorder alkaptonuria.

homologue. Chromosomes that are similar in structure, pair during meiosis, and correspond identically relative to their gene loci.

homozygous. The state in which an individual carries a pair of identical alleles at a particular locus.

hormone. A chemical substance synthesized in one organ of the body that influences functional activity in cells of other tissues and organs.

housekeeping genes. Genes that are theoretically expressed in all cells in order to provide the maintenance activities required by all cells.

H substance. A substance that appears to be an intermediate substrate in the synthesis of the blood antigens A and B. The gene involved is not allelic to the ABO locus.

human-leukocyte-associated antigens. Factors located on the surfaces of human leukocytes and concerned with the rejection or acceptance of foreign tissue; may also be important in the function of the immune system.

humulin. The name given to human insulin generated by bacteria that have the human insulin gene inserted into their genome by recombinant DNA techniques.

Huntington disease. A chronic progressive hereditary chorea characterized by irregular body movements, disturbance of speech, and gradual increasing dementia.

Hutchinson-Gilford syndrome. Same as progeria. A rare disorder characterized by a strikingly precocious aging, frequently causing death before ten or twelve years of age.

hybrid. The superiority of heterozygotes relative to the inbreds that formed them with regard to characteristics such as growth, development, and fertility.

hypertrichosis. A genetic trait in humans that consists of long hairs growing from unusual parts of the body.

ichthyosis. A congenital hereditary abnormality characterized by a dry, hard, scaly skin.

identical twins. See **monozygotic twins**.

immune system. The system that responds to the presence of antigens by producing antibodies and/or white blood cell activity.

inbred. A progeny that results from matings between relatives, and therefore is likely to have one or more situations in which two copies of a gene are identical by descent.

inbreeding. The mating of closely related individuals that results in increased homozygosity and uniformity of phenotypes among the individuals affected.

incestuous. Describes close consanguineous matings—that is, matings between relatives such as sibs or parent with progeny.

incomplete penetrance. The situation in which not all of the individuals having a certain genotype express the specific trait that usually results from that genotype.

independent assortment. The distribution of one gene pair of alleles at meiosis independently of other pairs of alleles.

inheritance. The transmission of genetic information from parents to their progeny.

insemination. The process of injecting semen into the vagina of a female.

insulin. A hormone manufactured by a small region of the pancreas and involved in the metabolism of carbohydrates.

intelligence quotient. A numerical assignment given to a person or a group of persons that is meant to indicate the mental age. The designation is based on the performance shown on standardized intelligence tests.

intermediary metabolism. The chemical reactions (metabolic steps) required to change precursor substances into products that are useful to the cell and the organism.

intersex. An individual with sexual characteristics intermediate between those of males and females; or individuals who may show secondary sex characteristics of both sexes.

intracytoplasmic sperm induction. Injection of a sperm cell nucleus into an oocyte, usually to overcome lack of sperm motility.

in utero. Within the uterus.

inversion. A chromosomal rearrangement such that an internal chromosome segment has been completely turned around end to end. As a result, the linear sequence of genes for that segment is reversed.

in vitro. Pertaining to experiments performed outside the organism—that is, inside glass vessels.

in vitro fertilization. The union of an egg and a spermatozoon under artificial circumstances—that is, within a glass vessel.

in vivo. Pertaining to experiments that are carried out such that the living organism is left intact.

ionizing radiation. Electromagnetic or corpuscular radiation that generates charged molecules in the material that it strikes.

IQ. See **intelligence quotient**.

karyotype. The composite view of somatic chromosomes of an individual arranged in a particular sequence by size and considered to be standard.

Kearns-Sayre syndrome. A neuromuscular disorder due to abnormalities in the DNA of mitochondria.

Klinefelter syndrome. A set of abnormal conditions in the phenotype of a human male that occasionally occurs due to the presence of more than one X (XXY).

Lamarckism. Inheritance of acquired characteristics proposed by French naturalist Jean-Baptiste Lamarck. Now believed to be largely erroneous.

laparoscope. A medical instrument for viewing inside the abdominal cavity. It consists of a long tube equipped with lenses.

LD$_{50}$. See **lethal dose 50**.

Lesch-Nyhan disease. A hereditary disorder that causes mental defects and spastic cerebral palsy, among other traits. The individual often has tendencies toward self-mutilation.

lethal dose 50. The radiation dose necessary to kill half of a population of organisms within a specified period of time.

lipofection. The technique of using liposomes to transfer genetic material into cells or organisms.

liposome. Small vesicles that form when insoluble fatty substances are mixed with water. These artificial bodies are being investigated as a possible means to transfer DNA into body cells.

locus. The specific position occupied by a given gene, or any of its alleles, on a chromosome.

Lysenkoism. The belief in the inheritance of acquired characteristics rather than the present gene concept.

male sterility. In plants, pollen abortion that occurs as a result of factors in the cytoplasm of cells that are maternally transmitted. These genetic factors only act in the absence of pollen-restorer genes.

malignancy. The condition in which tumor cells have the ability to grow progressively and often can kill their host.

manic depression. A mental disorder in persons that is characterized by periods of depression alternating with periods of excitement.

manifesting heterozygote. A female heterozygous for an X-linked recessive disorder who expresses the trait to approximately the degree as the affected males who are hemizygous for the disorder.

Marfan syndrome. An autosomal dominant genetic disorder in which affected persons are usually tall, long-limbed, loose jointed, and gaunt.

maternal effect. A nonlasting influence of the nuclear genes of the mother upon the phenotype of the immediate offspring.

maternal inheritance. The influence of the female on the phenotype of offspring controlled by either cytoplasmic or nuclear genetic factors.

meiosis. The cell division process whereby diploid precursor cells called germ cells form haploid sex cells.

melanin. A group of pigments ranging from black to brown to yellow that are found in the cells of skin, hair, the iris of the eye, other animal tissues, and some plants.

melanocyte. A pigment cell that contains melanin granules.

melanoma. A cancer composed of melanocytes.

Mendelian trait. Refers to Mendelian inheritance—that is, the concept of inheritance whereby traits are carried in cells and transmitted from one generation to another in the form of particles, now known to be genes.

messenger RNA. The ribonucleic acid product of gene action that serves as a chemical message to be translated into a specific polypeptide sequence.

metabolic block. A mutation that has the effect of preventing the synthesis of an enzyme that is necessary for some critical step in a metabolic pathway.

metabolism. The sum total of the various chemical reactions that are involved in the life functions of the living cell.

metastasis. The situation in which a disease, such as cancer, is transferred to another part of the body.

microsomal fraction. A subcellular cytoplasmic fraction of small particles that consists of ribosomes and broken fragments of the endoplasmic reticulum.

migration. Movement of one or more individuals from one geographic population to another that may result in a change in gene frequencies for either or both of the new populations.

millirem. 1/1,000th of a rem; a measure of radiation that takes into account damage done to tissue by different types of radiation, even though the dose may be the same.

mitochondrion. Small DNA-containing organelles in the cytoplasm of cells that are involved in many of the critical steps of cellular respiration.

mitosis. The process of nuclear division in which a replication of the chromosomes is followed by separation of the replication products and their incorporation into two daughter cells that are genetically identical to the original nucleus.

MN blood system. One of the blood systems in humans that involves two different antigens called M and N. The mode of inheritance of the blood types, MM, MN, NN, is one of codominance.

monosomic. Cytogenetically describes a diploid individual that lacks one chromosome of the normal set.

monozygotic twins. Same as identical twins. The members of the twin pair are genetically identical because they develop from one fertilized egg (zygote). At some time in early development, a separation occurs that results in two embryos.

mosaicism. The situation in which an organism is composed of cells of two or more different types with regard to genes and chromosomes.

multigenic. Same as polygenic. The mode of inheritance by which a trait is determined by two or more nonallelic pairs of genes.

multiple sclerosis. A disease in which the fatty myelin sheath around the spinal cord is damaged.

mutagen. A chemical or physical agent that significantly increases the rate of occurrence of mutations.

mutant. A cell or an organism that shows an observable change brought about by one or more mutations of the DNA.

mutation. The alteration of the genetic material. In the broad sense, mutation refers to any heritable change not due to segregation or genetic recombination.

myelogenous leukemia. Leukemia that is the result of disease in the bone marrow.

nanometer. One billionth of a meter.

natural selection. A differential reproduction of genotypes that comes about through the forces of natural processes favoring the individuals that are better adapted. This process has a general tendency to eliminate those unfit to survive.

neoplasm. An abnormal proliferation of cells in an organism.

neurofibromatosis. A condition in which multiple tumors are present in the skin and along nerves. Also called the elephant man disease.

neuron. A nerve cell.

nitrogenous base. A nitrogen-containing molecule with the properties of a base. The important bases in cells are the purines and pyrimidines found in the nucleic acids.

nondisjunction. The failure of either chromosomes or sister chromatids to separate normally during cell division, producing various types of aneuploid daughter cells.

nuclear transplantation. The technique used in cloning experiments. A nucleus from one source is placed into an enucleated egg from another source.

nuclein. The crude nucleoprotein complex initially isolated from nuclei by early biochemical investigators.

nucleotide. An organic compound composed of a purine or pyrimidine base, a pentose sugar, and a phosphate group.

nucleus. A membrane-bounded organelle of the cell containing the major part of the DNA of the cell—that is, the genetic information.

nullisomic. Pertains to a diploid cell or organism in which both members of a pair of chromosomes are absent.

offspring. The same as progeny—that is, the descendants of parents.

oncogene. The concept that genes, which are part of the normal chromosome complement, may become determinants for certain cancers.

oocyte. A cell that upon undergoing meiosis forms the ovum (egg).

overdominance. The condition whereby heterozygotes have a more extreme phenotype than either homozygote.

ozone. A colorless gas (O_3) forming a layer in the earth's atmosphere.

p53 gene. A tumor suppressor gene.

palindrome. DNA sequences that serve as recognition sites for restriction enzymes. The nucleotides of the sequences read the same on the complementary strands of the DNA molecule.

Parkinson disease. A disorder due to the loss of dopamine-producing brain cells. Symptoms are tremors, rigidity, stiffness, slowness of movement, and impaired balance.

parthenogenesis. The development of an embryo or an adult individual from an ovum (egg) without fertilization by a sperm.

Patau syndrome. A chromosome 13 trisomy condition in humans that causes a well-defined group of congenital defects.

pathogen. A disease-causing or toxin-producing organism.

pedigree. A diagram of two or more generations of kindred. The diagram uses symbols to depict the ancestral history of one or more traits for a given family.

phenotype. The actual appearance of discernible characteristics of an organism produced by its genotype interacting with the environment.

phenylalanine. One of the twenty amino acids used by organisms to synthesize polypeptides.

phenylketonuria. A recessive metabolic disorder characterized by an inability to oxidize the amino acid phenylalanine to tyrosine. If not treated early, the affected person will have severe mental defects.

phenylthiocarbamide. Known as PTC. A chemical taste, mostly bitter, to those who are carrying the dominant allele, but tasteless to those who are homozygous recessive for the trait.

Philadelphia chromosome. The designation given to the aberrant chromosome 22 in persons who have chronic myelogenous leukemia as a result of carrying the chromosome.

phocomelia. A condition in which an individual is born without arms. Also brought about through the use of thalidomide by pregnant women.

piebaldism. A dominant genetic disorder characterized by a phenotype of pigmented patches of skin and hair.

PKU. See **phenylketonuria**.

plasmid. Autonomously replicating elements found in the cytoplasm of bacteria. In most cases, plasmids are nonessential DNA entities.

pluripotent. Pertains to a cell or an embryonic tissue that has a large number of possible developmental fates because determination has not yet taken place.

polar body. One of the products of meiosis in the division of primary or secondary oocytes. Polar bodies are extruded as small bodies almost devoid of cytoplasm and they are nonfunctional in the reproductive process.

pollination. The transfer of pollen from the anther to the stigma within the same flower or from one flower to another.

polydactyly. The occurrence of more than the normal number of fingers and/or toes.

polygene. One of a group of genes that together controls the inheritance of a trait, usually a quantitative trait.

polygenic inheritance. The inheritance of quantitatively variable phenotypes. Traits usually depend on the interaction of numerous genes.

polymerase chain reaction. A technique in which cycles of amplification of a target DNA sequence can proceed to result in numerous copies of the sequence.

polymorphism. The existence of two or more genetically different classes in the same interbreeding population.

polynucleotide strand. A linear sequence of nucleotides chemically bonded together.

polypeptide. A chain of two or more amino acid residues linked covalently (peptide bonds) that will usually assume further structural configurations. One or more polypeptides will make up a protein.

polyploid. A cell or organism that has more than two complete sets of the basic haploid complement (genome) of chromosomes.

position effect. A phenotypic change that can occur in an organism when the responsible gene or genes change their positions relative to other genes in the genetic material.

postadaptation. A discredited idea that mutations can occur as a direct response to an environmental stress enabling the organism to adapt to that stress.

preadaptation. A new genetic characteristic arising from a mutation in an organism that often becomes a useful adaptation after an environmental change.

progenitor cell. A cell whose descendants can follow any of several developmental pathways.

progeny. Same as offspring. The individuals that result from a particular cross.

progeria. See **Hutchinson-Gilford syndrome**.

protein. Nitrogenous molecules composed of amino acids that form important structural and enzymatic constituents of the body.

proto-oncogenes. Genes that function in controlling the normal maintenance and proliferation of cells. They can also be converted into oncogenes by mutation and chromosome rearrangement.

pseudo-hermaphrodite. An individual with gonadal tissue of only one sex, but with some other features of both sexes.

PTC. See **phenylthiocarbamide**.

punctualistic model. A hypothetical model that describes evolutionary change as taking place with extensive branching of lineages at certain times—that is, speciation.

radon. A colorless radioactive gas formed by the decay of radium.

recessive. Any gene form (allele) that does not express or affect the phenotype when in the heterozygous state—that is, in the presence of another allele that masks it. Also describes any phenotype that will not manifest under these conditions.

reciprocal translocation. The mutual exchange of segments between nonhomologous chromosomes.

recombinant DNA. A DNA molecule that is made up of DNA from at least two different individuals, either of the same species or of different species.

recombination. The formation of combinations of alleles in the offspring that are not found in the parents. This can result either from a rearrangement following crossing over or from independent assortment.

regeneration. The regrowth of lost tissue or a whole part of an organism.

repair mechanism. A repertoire of enzymes and possibly other factors that give the cell the ability to repair some forms of genetic change.

repetitive DNA. Nucleotide sequences occurring repeatedly in chromosomal DNA.

replication. In the most commonly used sense, the synthesis of additional DNA from preexisting DNA by a process that involves the copying of templates in a precise manner.

restorer gene. A gene that reverses the changes brought about by cytoplasmic induced sterility.

restriction enzyme. One of a number of enzymes that internally cut DNA molecules at specific points in the molecule. These points are determined by a highly specific base pair sequence.

reverse mutation. A change in a mutant gene that restores its ability to produce a functional polypeptide.

Rh blood group. Antigens found on the red blood cells of some humans. People are classified as Rh+ or Rh−.

ribonucleic acid (RNA). A molecule that is usually single stranded, consisting of a chain of ribonucleotides in which the sugar is D-ribose. The commonly found nitrogenous bases of the molecule are adenine, cytosine, guanine, and uracil (rather than thymine as in DNA).

ring chromosome. A circular chromosome in higher forms of life due to an aberrant rearrangement.

RNA. See **ribonucleic acid**.

Salmonella **assay.** See **Ames test**.

schizophrenia. A form of mental disorder in which a personality split seems to take place; often a withdrawal from the usual human relationships accompanies the disorder.

scurvy. A deficiency disease resulting from a lack of ascorbic acid (vitamin C) in the diet.

segregation. The separation of the pairs of chromosomes, usually at meiosis, and the consequent separation of the alleles on these chromosomes.

selective medium. A medium containing ingredients such that only certain cells or microbes will survive on it. All other types would fail to show viability and growth.

self-fertilization. The formation of a zygote by fertilization involving gametes that are produced by a single individual. Many plant species can self-fertilize in this manner.

semen. The impregnating fluid from male animals that contains the spermatozoa.

seminiferous tubules. The mass of tubules that produce spermatozoa in the testes of an animal.

senescence. The process of aging.

sex chromatin. Refers to the Barr body. A chromatin mass in the nucleus of interphase cells of many mammalian species. The mass represents the X chromosome that is activated.

sex chromosome. Those chromosomes that are at least partly concerned with sex differentiation, and show a difference in number and/or morphology.

short tandem repeats. Repeats of two to ten DNA bases that are compared in DNA profiling.

sibs. The brothers and sisters from the same family.

sickle cell anemia. A severe hereditary disorder caused by a homozygous mutant gene that controls hemoglobin structure.

The red blood cells become sickle shaped under low oxygen tensions because of the defective hemoglobin.

sister chromatids. The two chromatids of the same chromosome after replication has occurred during cell division.

sociobiology. The discipline that seeks to establish that social behavior among animals has a biological basis.

somatic. Pertains to body cells as opposed to reproductive cells in the germ line. Normally, these cells have two sets of chromosomes, one from each parent.

somatic mutation. Genetic changes that occur in body cells—that is, those cells not destined to become gametes.

spermatozoon. A mature male gamete.

SRY gene. The sex-determining region on the Y. If the SRY gene is activated, the gonad develops into a testis; if not, an ovary forms under direction of other genes.

stem cell. A cell from which other cells arise and undergo differentiation.

surrogate mother. A substitute mother made possible by embryo transfer from one organism to another.

syndrome. A group of specific traits occurring together that are characteristic of a particular genetic condition, especially those considered abnormal.

Tay-Sachs disease. A type of genetic disorder that is inherited as an autosomal recessive and usually fatal at any early age. It is caused by an accumulation of gangliosides in the brain.

teratogenesis. The process by which gross congenital malformations are produced due to developmental errors.

testicular feminizaton. An abnormality in which a person with an XY genotype tends to have a female appearance with external female genitalia, but a blind vagina.

tetraploid. A polyploidy cell or organism that contains four complete sets of chromosomes, each of which is considered a genome.

thalidomide. A drug previously used as a tranquilizer and now determined to be highly teratogenic when taken by pregnant women.

totipotency. A property of a cell, or group of cells, such that they have the capacity to develop into a completely differentiated organism.

transcription. The synthesis of RNA from a DNA template by base pair complementarity.

transformation. The genetic modification of a cell or organism induced by the integration of purified DNA from cells of a different organism.

transgene. Genetic material that is experimentally transferred into organisms.

translocation. A structural chromosome aberration in which a portion of one chromosome breaks away and becomes incorporated at a different site, either within the same chromosome or within a nonhomologous chromosome.

transplantation. Transfer of an organ or a part of an organism to another organism, or to another location within the same organism.

transposon. A class of genetic elements that can move about within the genome and renders target genes unstable.

transsexualism. Refers to voluntary sex changes usually brought about by surgery and hormonal applications.

triploid. Describes a polyploidy cell or organism having three sets of the basic haploid chromosome complement—that is, genomes.

trisomic. Describes a condition in which a cell or an individual has one extra chromosome; hence, three homologues exist for one of the chromosomes of the basic set rather than two.

trisomy-13. Also known as Patau syndrome. A trisomic condition in which chromosome 13 is represented three times. The infant is born with severe internal and external anomalies and mental defects.

trisomy-18. A trisomic condition in which chromosome 18 is represented three times. The defect results in a child with multiple congenital malformations. The disorder is also called Edwards syndrome.

trisomy-21. A genetic abnormality in humans in which chromosome 21 is represented three times. The disorder is also called Down syndrome.

tumor. A local mass on or in any part of the body. It arises by an abnormal growth of the tissues. It may or may not be malignant.

Turner syndrome. Also called XO. In humans, an abnormal condition in which the person is monosomic for the X chromosome and lacks an accompanying Y chromosome. The affected persons are sterile females.

virus. A submicroscopic organism that is a noncellular obligate parasite composed of a nucleic acid and a protein shell.

Werner syndrome. A syndrome in humans in which the symptoms of aging are manifested prematurely, usually in early adulthood.

xeroderma pigmentosum. A skin disease characterized by the development of numerous pigmented spots. Frequently the condition develops into skin cancer.

XO syndrome. See **Turner syndrome**.

XXY karyotype. See **Klinefelter syndrome**.

XYY syndrome. The presence of an extra X chromosome, resulting in males who are usually taller than average and with a phenotype similar to that of an individual with Klinefelter syndrome.

Zea mays. Scientific species name for corn. Not only a valuable crop plant, but also an often used model organism for genetic research.

zygote. Usually a diploid cell formed by the fusion of two haploid gametes during fertilization. In higher forms of life, it is the fertilized egg before cleavage.

zygote intrafallopian transfer. An assisted reproductive technology in which an ovum fertilized in vitro is placed in a woman's uterine tube.

INDEX